Potatoes

WINROCK DEVELOPMENT-ORIENTED LITERATURE SERIES

Steven A. Breth, series editor

Potatoes: Production, Marketing, and Programs for Developing Countries was prepared under the auspices of Winrock International Institute for Agricultural Development, the International Potato Center, and the International Food Policy Research Institute.

ALSO IN THIS SERIES

Rice in the Tropics: A Guide to the Development of National Programs, Robert F. Chandler, Jr.

Small Farm Development: Understanding and Improving Farming Systems in the Humid Tropics, Richard R. Harwood

Successful Seed Programs: A Planning and Management Guide, Johnson E. Douglas

Tomatoes in the Tropics, Ruben L. Villareal

Wheat in the Third World, Haldore Hanson, Norman E. Borlaug, and R. Glenn Anderson

Cassava: New Potential for a Neglected Crop, James H. Cock

Potatoes

Production, Marketing, and Programs for Developing Countries

Douglas Horton

Westview Press (Boulder)
IT Publications (London)
1987

Winrock Development-Oriented Literature Series

Published in 1987 in the United States of America by Westview Press, Inc.; Frederick A. Praeger, Publisher; 5500 Central Avenue, Boulder, Colorado 80301

Published in 1987 in the United Kingdom by Intermediate Technology Publications, 9 King Street, London, WC2E 8HW, England

Library of Congress Cataloging-in-Publication Data
Horton, Douglas E.
Potatoes : production, marketing, and programs in developing countries.
(Winrock development-oriented literature series)
Bibliography: p.
Includes index.
1. Potatoes—Developing countries. 2. Potato industry—Developing countries. 3. Potatoes—Government policy—Developing countries. I. Title. II. Series.
SB211.P8H68 1987 338.1′73491′091724 86-28304
ISBN (U.S.) 0-8133-7197-X

ISBN (U.K.) 0-946688-09-5

Printed in the United States of America

The paper used in this publication meets the requirements of the American National Standard for Permanence of Paper for Printed Library Materials Z39.48-1984.

10 9 8 7 6 5 4 3 2 1

Contents

Tables and Figures

Tables

Figures

Foreword

In the next three decades, the Third World will experience the most dramatic increases in demand for food in human history. This explosion derives from rapid population growth, especially in Africa, and from rising per capita income in much of the Third World. Agriculture will also remain the primary source of income increase for most people. If food production is to keep pace with the demand and be a dynamic source of income growth, scientists and policymakers must develop new and creative approaches to agricultural development. Concentration on cereals must give way to a better understanding and use of a range of food sources. Simultaneously, institutions and strategies must be developed that will facilitate efficient exploitation of those sources. The potato offers a particularly significant potential for increasing food production and incomes in developing countries. However, realizing the potato's potential role in agricultural development requires that a number of complex institutional problems be overcome.

Dr. Horton addresses these issues by setting out to reach individuals who are concerned with food in developing countries but who have little background in either economics or potato production. Economics is important because fundamental realities of supply, demand, and markets will determine how and by whom foods will be grown and used. Potatoes are important because they are a major food crop whose production is increasing rapidly. Dr. Horton was able to combine a long period on the staff of the International Potato Center with a year at the International Food Policy Research Institute (IFPRI) so as to make a synthesis of the technical and economic aspects of the problem.

It is no simple task to boil down complex research findings into clear, straightforward statements. But this must be done so that administrators and research managers can capture the essentials without wading through the abundance of sometimes conflicting information that is now available. This is especially true for potatoes because our knowledge of the crop and the principal constraints to production and use in developing countries was very limited just 10 years ago, but is now growing rapidly. In this book, Dr. Horton has synthesized in a clear manner a vast amount of information on potato research and economics for the benefit of the nonspecialist. This model could probably serve very well for many other food crops in developing countries.

Dr. Horton's book reflects the spirit of the International Potato Center (Centro Internacional de la Papa or CIP) and its accomplishments through collaboration with many institutions in a modest, cost-effective way. It is gratifying to see that from CIP's simple beginnings in 1971, the potato is finally being recognized as an important crop by scientists and policymakers and that potato programs are now benefiting producers and consumers in developing countries. We believe this book will prove to be very useful for all of us concerned with alleviating food shortages and raising incomes in poor countries.

J. Mellor
Director, IFPRI
Washington, D.C., U.S.A.

R. L. Sawyer
Director-General, CIP
Lima, Peru

Acknowledgments

I am indebted to Winrock International for inviting me to write this manuscript and to Steven Breth for his encouragement and professional contributions throughout the book's preparation. I am also grateful to the International Potato Center, and particularly Richard L. Sawyer, for providing me with what turned out to be something more than a sabbatical year for completing the book. The International Food Policy Research Institute generously furnished a congenial and stimulating working environment and indispensable assistance in processing a massive amount of data as well as the text. Many present and former colleagues at CIP have kindly shared with me published and unpublished materials on their work and have made useful comments without which I could not have written this book. I am especially indebted to W. G. Burton and O. T. Page on technical aspects of the potato; Humberto Mendoza and Carlos Ochoa on breeding and taxonomy; Susan Poats and Jennifer Woolfe on consumption and nutrition; Efrain Franco, Enrique Mayer, and especially Robert Rhoades on potato production zones and systems; Anibal Monares and James Bryan on seed production and distribution; Greg Scott on marketing; Robert Booth and Roy Shaw on post-harvest technology; Primo Accatino and Kenneth Brown on regional and national programs; Manuel Pina on training and communications; and Gelia Castillo and Richard Sawyer on the institutional strategies that have made CIP unique. Lyle Sikka provided me with extensive information on Bangladesh, and Angelique Haugerud and Michael Potts supplied useful information and comments on the sections dealing with Rwanda and the Philippines. Peter Ewell gave insightful comments on the entire manuscript.

I am also indebted to my colleagues and friends at IFPRI. John Mellor made my stay possible, encouraged my work, and commented thoughtfully on the original manuscript. Leonardo Paulino, Loraine Halsey, and Jackie Gilpin handled administrative matters with great efficiency and courtesy. Tricia Klosky furnished excellent library support and warm hospitality. Bruce Stone generously shared his extensive knowledge of Chinese agriculture. Hannan Ezekiel and Raisuddin Ahmed made useful comments on sections of the manuscript dealing with India and Bangladesh and encouraged me to expand the coverage of the potato's export potential. Frances Walther leavened the environment with an occasional potato joke. Mustapha Rechache and Richard Schuman supplied top-notch data and text processing often under severe time constraints.

The entire original manuscript was reviewed by Richard Harwood, W. O. Jones, Eduardo Venezian, and a fourth reviewer who wishes to remain anonymous. Date van der Zaag and Henk Beukema commented on Chapters 2, 6, and 7. Together they provided over ninety pages of typewritten comments (an average of one page of comments for each four pages of manuscript!). As might be expected, the reviewers concurred on some topics, but on others they disagreed sharply. I have attempted to incorporate many of their suggestions into the final version of the book. I am particularly indebted to W. O. Jones for his comments on style and tone and suggestions for improving Chapters 1, 4, 6, and 7; to Richard Harwood for his many comments on both technical and socioeconomic topics and useful suggestions for improving Chapter 6; and to Eduardo Venezian for suggesting improvements in the treatment of economic topics in several chapters. I, of course, assume full responsibility for any remaining errors and omissions.

Doug Horton
Washington, D.C.

Introduction

The potato is one of the most important food crops in developing countries as well as in developed countries. Due to its importance in Europe as a cheap food, livestock feed, and source of starch and alcohol during the industrial revolution, the potato is most often thought of as a crop whose production and use are largely confined to industrialized nations. In fact, developing countries today produce about 30 percent of the world's potatoes, and production in those countries is expanding more rapidly than that of most other food crops. As a result, potato cultivation is becoming an increasingly important source of rural employment, income, and food for growing populations. In monetary terms, potatoes are now the fourth most important of the developing world's food crops.

Policies affecting the potato crop are becoming increasingly important in developing areas. Unfortunately, the information required for sound policy decisions is often not readily available because it is scattered through numerous text books, scientific papers, statistical reports, and unpublished documents. In addition, much of the information available is out of date or misleading because it was written with industrial nations in mind. Because managers of development programs may be unclear about differences between the food systems and technology needs of rich and poor countries, these programs often attempt to "transfer" inappropriate technology to developing countries. A better understanding of the present and potential role of potatoes within the food systems of developing countries can help researchers, policymakers, and development agencies avoid these problems in the future.

This book has two major goals. The first is to inform readers about the essential aspects of potatoes in the food systems of developing countries. This includes recent trends in potato production and use, biology of the crop, the policy environment, production zones and systems, supply and demand, marketing problems, consumption patterns, and nutritional value. Information on these topics is presented in Chapters 1 through 6. The second goal is to outline the major issues that need to be considered in setting priorities and for implementing and assessing the impact of potato programs. These issues are discussed in Chapters 7 and 8. These eight chapters address a number of broad questions to which policymakers and researchers concerned with potatoes in developing countries need answers.

1. How important are potatoes in developing countries, and what are the recent trends in potato production and use?
2. What biological or physical features make potatoes special?
3. Who grows potatoes in developing countries, where, and how?
4. Who eats potatoes and in what quantities?
5. What nutrients do potatoes provide in the human diet?
6. How do price and income changes affect the demand for potatoes?
7. How can price controls, storage, processing, and foreign trade stabilize prices or expand markets for potatoes?
8. Why should scarce public resources be allocated to potato research and extension in developing countries?
9. What are the essential ingredients of a successful potato program?
10. What impact have successful potato programs had?
11. What are the future prospects for potatoes in developing countries?

Although the subject of this book is potatoes, the approach used, the topics covered, and many of the issues raised apply to other food crops as well. I hope this book will stimulate researchers and policymakers to take a closer look at potatoes and other "minor crops" within the food systems of their countries.

1
World patterns and trends in potato production and use

Among the world's food crops, the potato ranks in the top five in tonnage, and as a source of food energy it provides about one-sixth as many calories as rice, wheat, and maize (Table 1).

The potato (*Solanum tuberosum*) has its origin in the high Andes of South America, but for the last two centuries most potatoes have been grown in Europe. For this reason, many people think of it as a European crop—the "Irish" potato—and assume that world trends in potato production and use parallel European trends. This is not the case.

Since World War II, world potato production has grown by less than half a percent per year. However, this lackluster global trend masks strongly divergent regional and national changes. In Western Europe, average potato production has dropped by 1.5 percent a year, while in Eastern Europe it has been flat. Elsewhere, production has increased: in North America and Oceania by 1 to 2 percent per year; in Latin America by nearly 3 percent per year; and in Africa and Asia by 4 percent per year. In the developing countries, total potato production has more than doubled since 1965 (Fig. 1).

Origin and spread of the potato crop

South American origin

The potato was first cultivated in the Andes in the vicinity of Lake Titicaca near the present border of Peru and Bolivia.

TABLE 1
World Production of Major Food Crops, 1984

Crops	Production (million tons)	Dry matter (million tons)	Edible portion	
			Energy (trillion kcal)	Protein (million tons)
Cereals				
Wheat	522	456	1,320	52.7
Rice, paddy	470	414	1,143	21.1
Barley	172	153	419	9.9
Maize	449	387	1,376	35.2
Rye	31	28	89	3.2
Oats	43	40	76	2.8
Millet	31	27	95	2.9
Sorghum	72	64	226	7.5
Roots and Tubers				
Potatoes	312	63	192	5.3
Sweet potatoes	117	34	108	1.6
Cassava	129	52	110	0.5
Yams	25	7	22	0.5
Cocoyams	6	2	5	0.1
Pulses				
Beans, dry	15	14	53	3.4
Broad-beans, dry	4	4	14	1.0
Peas, dry	11	9	36	2.6
Chickpeas	7	6	23	1.3
Lentils	2	1	5	0.4
Oilseeds				
Soybeans	90	81	362	30.7
Groundnuts in shell	21	19	85	3.9
Vegetables				
Cabbages	38	3	7	0.4
Tomatoes	59	4	11	0.6
Onions, dry	23	3	8	0.3
Carrots	12	1	4	0.1
Bananas & plantains	61	21	53	0.5

Source: FAO, *Production yearbook 1984* (Rome, 1985); and USDA, *Composition of foods* (Washington, D.C., 1975).

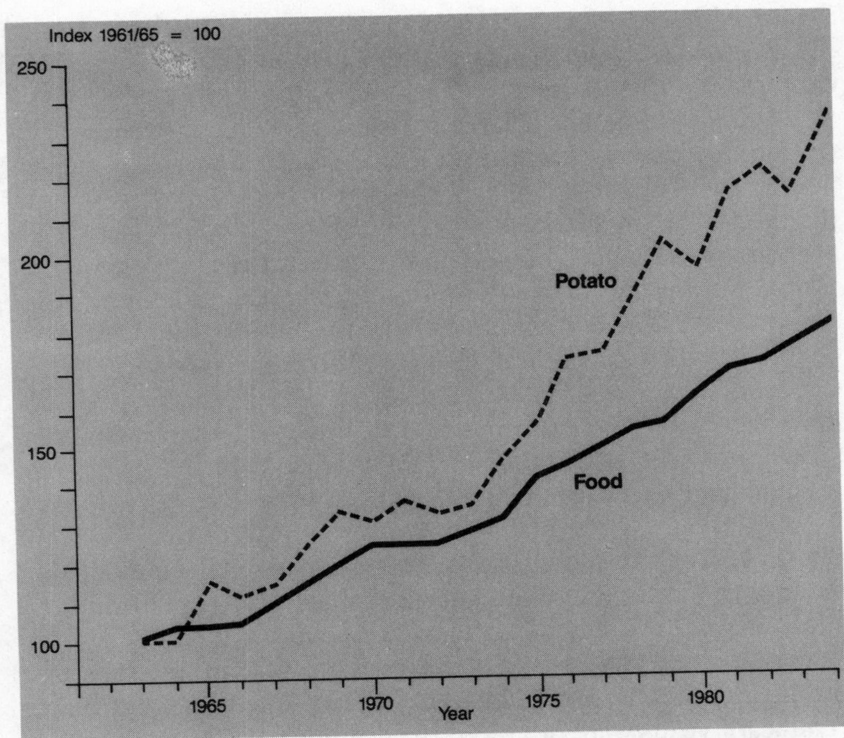

Figure 1. Trends (3-year moving averages) in potato production and total food production in developing countries. *Source:* FAO, *Production yearbook 1976; 1984* (Rome, 1977; 1985). Unpublished data provided by FAO Basic Data Unit.

Once domesticated, the potato crop spread throughout the Andes, and by the time of the Spanish conquest in the early sixteenth century, farmers were cultivating hundreds of varieties throughout the highland areas of what are now Bolivia, Chile, Colombia, Ecuador, and Peru.

Introduction of the potato in Europe

Potatoes were sold in Seville as early as 1573. They were brought there by Spanish sailors. From Spain, the potato spread throughout Europe, first as a botanical curiosity, later as a food crop. Legend has associated Sir Walter Raleigh and Sir Francis

Drake with the introduction of potatoes into England, but it seems more likely that potatoes first arrived on a Spanish ship captured by English seamen around 1590.

There has been considerable controversy as to which subspecies of potato was first brought to Europe and from what part of South America: *Solanum tuberosum* subspecies *andigena* from the Andes or *Solanum tuberosum* subspecies *tuberosum* from southern Chile. The *andigena* subspecies tuberizes (produces tubers) under the 12-hour days of tropical latitudes, but does not tuberize in Europe until late in the season—September or October—when the natural daylength shortens to about 12 hours. Consequently, Soviet botanists believed this subspecies could not have been the one first introduced in Europe. Experiments have shown, however, that the potato can evolve from one subspecies to the other in a few generations of selection for earliness. Leading authorities now believe that the potato came to Europe from the Andes and, through natural selection, evolved from the parent subspecies (*andigena*) to the derived subspecies (*tuberosum*) by 1750, when Linnaeus described the European potato.

Because the Andean potatoes brought to Europe tuberized late in the season, the plants were often killed by frost before significant yields were produced. Andean potatoes flourished only in a few places, like southern Ireland and France, where winter is mild. Much later, as natural selection produced earlier-maturing, higher-yielding *S. tuberosum* varieties, potato cultivation became economically feasible elsewhere in Europe.

In Europe potatoes were fed to livestock long before they became a staple in human diets. People rejected potatoes as being unclean, unhealthy, or even poisonous. Physicians warned that eating potatoes could result in ills ranging from indigestion and flatulence to rickets and syphilis. On the grounds that potato consumption caused leprosy, a French provincial parliament banned potato cultivation altogether. In the eighteenth century, however, the potato began to gain respectability in continental Europe when statesmen and peasants alike began to realize that in times of lean harvests this "exotic" vegetable could be substituted for ordinary food.

The potato in Ireland

Perhaps no single crop and no country have been so closely associated in modern history as have the potato and Ireland. For over 200 years, the potato was Ireland's most important crop and practically the sole food of the poor. The climate of southern Ireland is ideal for potato production, and by the mid-1700s, the potato was already producing more food calories per hectare than any other crop. Normal yields in Ireland in the early 1800s were over 10 tons per hectare—just about the average for all developing countries today. The potato's ascendance as a staple food resulted largely from social conditions in Ireland. During the sixteenth century, unrest and warfare were chronic, and cattle, food stocks, and standing crops were often confiscated or destroyed; potatoes that were still in the ground escaped. Farmers, realizing this, did not harvest and store their potatoes, but dug them up as required. Tubers left in the ground served as seed for the next crop.

In the seventeenth and eighteenth centuries, absentee landowners rented their estates to middlemen, who broke up the land into small rental units. The subletting system, high population density, and poverty led to a minute fractioning of landholdings. The comparatively large amount of food yielded by the potato per unit area, coupled with dire necessity of the farmer, rendered potato cultivation almost obligatory. Dependence upon the potato as a food became so great that in 1780, average daily potato consumption was reported to be over 3 kilograms per person.

Not only was Ireland's food supply highly dependent upon a single crop—the potato—it was dependent upon thousands of hectares of potatoes that had all evolved from a few tubers brought from South America. The narrow genetic background of the potatoes grown in Ireland made them extremely vulnerable to disease attack. If one plant was susceptible to a disease, most others were likely to be susceptible too.

In the 1840s an outbreak of late blight ravaged potatoes. This fungal disease destroys stems and foliage. The ensuing "potato famine" was the worst European disaster since the Black Death 500 years before. A million people died and 1.5 million emigrated, out of a total population of around 8 million. After the famine, Ireland began to diversify its agricultural economy, and the area under potato cultivation has tended to decline.

Spread of potatoes in Europe

From Spain, the potato spread to continental Europe; from England, it was dispersed throughout the British Isles and to parts of northern Europe. By 1600, potatoes were sent from Spain to Italy and from there to Germany, and within a hundred years, they had reached most of German-speaking Europe, although they were not widely cultivated until the 1800s. Potatoes also arrived in France by about 1600, and they were an important livestock feed in southern France 50 years later. The potato was introduced into Russia—now the world's largest producer—before 1700, but insignificant amounts were grown until about 1800.

Only when potatoes became a low-cost source of food energy (due primarily to the selection of adapted varieties) and when other foods became scarce (primarily as a result of crop failures, wars, or famines) did Europeans begin to replace their traditional staples such as buckwheat and oats with potatoes. The popularity of the potato in Europe reached its zenith around 1850: It was the least expensive source of food energy consumed in Europe, the most important food crop after wheat, a major livestock feed, and the principal source of starch and alcohol. Since then, per capita potato production and use has declined in Europe, as other foods, feeds, and sources of starch and alcohol have become cheaper, and as consumers have diversified their diets, reducing dependence on low-cost staples.

Spread of potatoes to other parts of the world

Potatoes reached most other parts of the world through the European colonial powers, rather than directly from South America. North America first received potatoes from England via Bermuda in 1621; there is no record of an introduction from South America until the 1860s.

Travelers carried potatoes to the tropics and subtropics. British missionaries, for example, took potatoes to many parts of Asia in the seventeenth century, and Belgian missionaries introduced them to the Congo in the nineteenth century. But, until recently, use of the potato as a food crop in the tropics was severely limited by difficulties inherent in producing and storing them in the lowlands.

The potato arrived in China around 1700 from the Dutch East Indies (now Indonesia). Introductions were also made to northern China from Russia. Potato production spread throughout the country and is now particularly important in the northeast (Heilungkiang), Inner Mongolia, and in steep mountain valleys in the southern provinces.

The potato was brought to India in the 1600s by Portuguese traders who landed north of Bombay. Prior to 1700, potatoes were grown as a garden vegetable in parts of western India, but the potato reached southern India only in the 1880s. The British promoted potato cultivation in the hill areas of northern India. By the 1820s, farming households were growing potatoes in garden plots in the Simla hills—the present site of India's Central Potato Research Institute. Within a few decades, potatoes had become a staple food in some remote hill areas. From the hills, potatoes spread to the plains, and by 1900, small plots of potatoes were found near towns scattered throughout northern India. Potato cultivation spread on the plains in the winter season as seed systems emerged, irrigation and storage methods improved, population grew, and demand for food increased.

Potatoes were introduced into Tibet from both China and India. By 1900 the chief dish of the Tibetans was reported to be a stew of meat and potatoes, turnips, cabbage, and other vegetables accompanied with dried cheese. By that time, potato cultivation was also common in Bhutan, Nepal, Sikkim, and Kashmir.

The potato was first carried to Japan around 1600 by Dutch traders from Java. The Japanese did not care for its taste, but after a succession of floods and famines in the late seventeenth century, they began expanding its cultivation in cooler, less fertile areas. It is not known when the potato reached Korea, but Koreans have long grown the crop, and in some mountainous areas, potatoes are an important livestock feed and dietary staple.

Potatoes were introduced into Persia in the late eighteenth century. About 50 years later, they reached Syria, where they were known as "colocasia of the foreigners." Late in the nineteenth century, the Turks promoted potato cultivation in the plain of Erzerum, which is still a major production zone.

In sub-Saharan Africa, introduction of the potato followed colonization. And just as the Malayans called the potato the Dutch, English, or Bengal yam, in central Africa it was termed the "white man's yam" or "European root." After seeing Christian missionaries growing potatoes in gardens, natives in some localities began growing them for sale to Europeans. Generally, however, Africans began eating potatoes only when forced to by famine or government decree. In Rwanda, for example, Belgian colonists introduced potatoes in the late 1800s, but the natives considered them taboo. Only when famine struck did people begin eating potatoes.

Although the evidence is sketchy, it seems that, as in Europe, the spread of potato production and consumption in what is now called the developing world was strongly influenced by the suitability of the environment for the crop, development of production and post-harvest systems that were appropriate for specific environments, and food habits and needs. Potato cultivation expanded first in areas with cool, rainy climates. Often local food habits and myths discouraged potato consumption long after its introduction, and until times of food shortage, most potatoes were eaten primarily by European colonists rather than by the indigenous population. In warm environments for potato production—such as the Indo-Gangetic plain—potatoes were restricted to garden plots until quite recently when a series of new supply and demand factors induced the spread of potato production and consumption.

Recent trends in potato production and use

Published data on potato production and use in developing countries are not very reliable. Government statisticians usually give highest priority to compiling information on commodities like cereals, rubber, and coffee, which are traded on international markets. Domestic food crops receive less attention. Statisticians also have difficulty estimating the production of crops like potatoes that are often grown in isolated areas on small, irregular plots. And computing the amounts of potatoes that are used for seed, livestock feed, processing, and human consumption is even more demanding. In food balance sheets, human consumption is generally estimated as a residual, after subtracting other uses and waste from the total domestic availability. Unfortunately, little research has been done to quantify the various uses of potatoes in developing areas. Seeding rates, for example, are usually assumed to be near the rates prevailing in developed countries. However, farmers in many developing areas actually use lower rates. Estimates of waste and the amounts of potatoes that are fed to livestock and processed are also more often based on popular notions than on fieldwork.

The data on potato production and use in this chapter are based on Food and Agriculture Organization (FAO) estimates, adjusted because of known underestimation in the past. As these figures are averages, they do not reflect differences between ecological zones and rural and urban areas within countries. Chapters 5 and 6 discuss some of the important differences in potato production systems and consumption patterns observed within developing countries. An urgent task for researchers in every country is to improve the database on national potato production and use. Such information can be invaluable in planning, implementing, and evaluating the impact of research and extension programs.

Production

World potato production was about 135 million tons at the turn of the century, 250 million tons around 1950, and in the mid-1980s is about 290 million tons. During the first half of the century, Europe (including the USSR) produced about 90 percent of the world's potatoes. After World War II, potato production began to fall in Europe. Since 1960 potato production in Western Europe has dropped by more than a third (Table 2). Potato production also shrank by about 10 percent in Eastern Europe and the USSR, but elsewhere it grew. For the developing market economies as a group, potato production has grown at a faster pace than production of most other food crops (Fig. 2).

TABLE 2
Percentage Change in Potato Production and Consumption in Major Regions and Selected Countries Between 1961/65 and 1981/83

	Percentage change in			
	Area	Production	Yield	Per capita consumption
World	−15	1	18	−15
North America	−8	25	35	14
Canada	−6	27	35	−2
USA	−8	24	35	16
Western Europe	−48	−36	24	−13
France	−73	−53	73	−25
Germany (FRG)	−71	−67	15	−33
Italy	−59	−28	77	−4
Netherlands	28	60	25	−11
Spain	−14	17	36	6
United Kingdom	−40	−14	43	1
Yugoslavia	−9	−2	8	−11
Eastern Europe & USSR	−23	−10	18	−21
Czechoslovakia	−60	−38	53	−31
Germany (GDR)	−34	−26	13	2
Hungary	−66	−24	125	−37
Poland	−22	−17	6	−8
Romania	−9	99	120	8
USSR	−21	−5	20	−25
Sub-Saharan Africa	90	146	29	40
Kenya	−26	39	88	−8
Madagascar	156	164	3	29
Rwanda	93	271	93	169
South Africa	59	145	54	27
Latin America	0	48	48	−5
Argentina	−39	17	91	2
Brazil	−13	66	92	23
Chile	−8	6	16	−42
Colombia	139	180	17	78
Cuba	79	182	57	47
Mexico	49	147	65	36
Peru	−25	5	40	−26

TABLE 2, cont.

	Percentage change in			
	Area	Production	Yield	Per capita consumption
North Africa & Near East	85	134	26	53
Algeria	236	159	−23	57
Egypt	190	189	0	103
Morocco	68	120	31	94
Syria	375	681	64	207
Turkey	26	92	53	25
Asia	29	81	40	22
Bangladesh	96	202	54	n.d.
China	25	73	39	16
India	87	234	79	123
Japan	−40	−7	56	−6
Korea (DPR)	−9	70	86	1
Korea (ROK)	−29	8	51	−52
Nepal	35	25	−7	−1
Pakistan	200	258	19	94
Oceania	−14	36	58	9
Australia	−8	58	72	15

n.d. = no data

Source: FAO, *Production yearbook 1976; 1983* (Rome, 1977; 1984); FAO, *Food balance sheets 1964–1966 average* (Rome, 1971) and FAO, *Food balance sheets 1979–1981 average* (Rome, 1984).

Of total world production, Eastern Europe and the USSR produce just under half (Table 3), and Western Europe produces about 15 percent. Nearly a quarter of the world's potatoes are grown in the Far East (including China), 10 percent are produced in the Americas and Oceania, and less than 5 percent are cultivated in Africa and the Middle East.

Germany, once the world's leading potato producer, has registered a precipitous decline in production. Since 1960, potato production in West Germany has fallen by two-thirds, and per capita production by a bit more. Potato production has also dropped by more than half in France. In contrast, potato

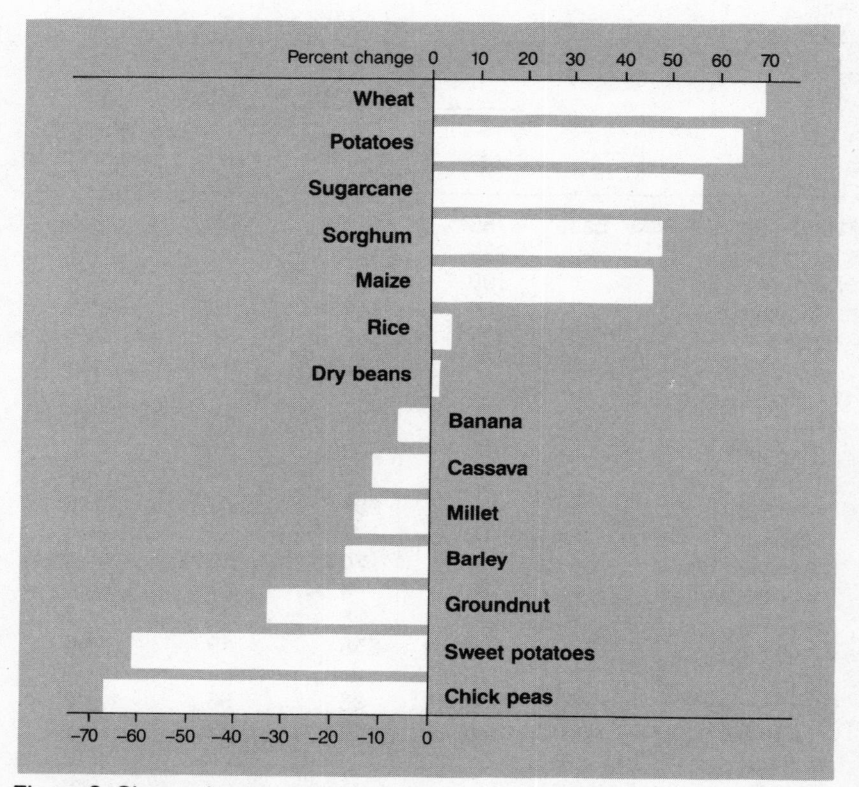

Figure 2. Change in per capita production of major food crops in developing market economies, 1961/65–1982/84. *Source:* FAO, *Production yearbook 1976; 1984* (Rome, 1977; 1985).

production has tripled since 1960 in Rwanda, Bangladesh, India, and Pakistan and more than doubled in Madagascar, Cuba, Mexico, and several North African and Middle Eastern countries. As a result of these divergent trends, more potatoes are now grown in Asia than in Western Europe, North America, and Oceania combined. China's production alone equals that of Western Europe, and India produces more than West Germany.

In South America, where potatoes have long been a staple food, potato production has grown less rapidly, in part because consumers are diversifying their diets and moving away from traditional staples. Agricultural and trade policies have also discouraged potato production in some countries, like Peru.

TABLE 3
Potato Production and Consumption in Major Regions and Selected Countries, 1981/83

	Area (000 ha)	Production (million tons)	Yield (t/ha)	Per capita production (kg)	Per capita consumption (kg)
World	20,281	286.1	14	69	33
North America	618	18.0	29	70	55
Canada	112	2.7	24	108	79
USA	506	15.4	30	66	52
Western Europe	2,055	46.3	23	124	79
France	208	6.2	30	114	78
Germany (FRG)	263	7.4	28	121	78
Italy	150	2.8	19	48	41
Netherlands	165	6.0	37	422	82
Spain	339	5.3	16	139	112
United Kingdom	193	6.3	33	112	103
Yugoslavia	282	2.7	9	117	57
Eastern Europe	10,185	133.8	13	351	103
Czechoslovakia	197	3.5	18	227	77
Germany (GDR)	478	8.9	19	534	142
Hungary	86	1.5	18	142	57
Poland	2,219	36.3	16	1003	118
Romania	285	5.2	18	231	71
USSR	6,866	77.8	11	288	107
Sub-Saharan Africa	391	2.9	7	13	9
Kenya	38	0.3	7	15	11
Madagascar	31	0.2	7	25	16
Rwanda	35	0.2	7	45	33
South Africa	70	0.9	13	30	20
Latin America	1,044	11.2	11	31	24
Argentina	109	2.0	19	73	68
Brazil	174	2.0	11	15	12
Chile	83	0.8	10	74	42
Colombia	160	2.0	13	76	54
Cuba	14	0.3	18	25	24
Mexico	69	0.9	13	12	10
Peru	192	1.6	8	84	65

TABLE 3, cont.

	Area (000 ha)	Production (million tons)	Yield (t/ha)	Per capita production (kg)	Per capita consumption (kg)
North Africa & Near East	564	7.7	14	30	29
Algeria	81	0.6	7	29	29
Egypt	70	1.2	16	26	19
Morocco	35	0.5	13	21	15
Syria	19	0.3	16	31	24
Turkey	180	3.0	17	64	50
Asia	5,378	65.2	12	27	16
Bangladesh	106	1.1	10	11	9
China	4,005	46.7	12	46	26
India	748	9.9	13	14	10
Japan	128	3.5	27	30	15
Korea (DPR)	132	1.6	12	86	62
Korea (ROK)	36	0.5	15	13	7
Nepal	57	0.3	6	22	14
Pakistan	45	0.5	11	5	4
Oceania	45	1.1	24	60	51
Australia	37	0.9	24	59	50

Source: FAO, *Production yearbook 1983* (Rome, 1984); and FAO, *Food balance sheets 1979–1981 average* (Rome, 1984).

Potato yields have risen throughout the world, most notably in developing regions. In Europe, since World War II, increases in potato yields have not kept pace with increases in cereal yields, reflecting greater improvement in cereal varieties and production systems. As a result, potatoes have become more costly than other energy sources, and use of potatoes for livestock feeding and processing has fallen. Demand for fresh potatoes has also dropped as diets have changed. For these reasons, European farmers have reduced the area and production of potatoes.

In developing areas, with the exception of Asia, potato yields have increased at least as much as cereal yields and much more than yields of other root crops. Improvements in varieties, seed systems, and post-harvest technology have helped bring down the cost of producing potatoes, improving their competitive position on farms in many developing areas. Population growth, rising incomes, and changing food habits have also stimulated potato production.

The balance of world potato production is gradually shifting from the developed to the developing countries and from the temperate to the tropical and subtropical zones. Only about 60 percent of the world's potatoes are now grown in Europe; 10 percent are grown in other developed countries, and 30 percent in developing countries.

Consumption

Consumption trends do not always parallel the trends in production, for potatoes are also fed to animals and serve as the raw material for such industrial products as starch and alcohol. A rather large part of the harvest is also used for seed (Table 4). In Western Europe, as a result of declining feed and industrial use of potatoes, per capita potato production has fallen by more than 40 percent since 1960, but per capita potato consumption has dropped by less than 15 percent. The share of the crop that is wasted and used as seed has also fallen somewhat.

Use of potatoes in livestock feeding and industry has declined as potatoes have become more expensive relative to alternative feedstuffs and energy sources. Prior to World War II, nearly half the potatoes grown in Western Europe were fed to livestock, but now less than 20 percent are. Hogs and potatoes were frequently raised together on small farms, but now they are produced separately on large, specialized farms in different areas. Swine producers find it more profitable to feed concentrated rations of barley, cassava chips, and soybeans than to feed boiled or ensiled potatoes that would have to be trucked in from other areas. Use of potatoes as a raw material for industry has also declined in all but a few countries.

In developing countries, where few potatoes are processed or fed to livestock, production and consumption trends are more

TABLE 4
National Food Balance Sheets for Potatoes, 1979/81

	Sources (million tons)			Uses (% of domestic availability)						
	Production	Net imports and changes in stocks	Total domestic availability	Seed	Feed	Industry	Waste	Food	Consumption (kg per capita)	Food calories per capita
U.S.	14.88	0.16	15.04	7	2	5	8	78	51	98
France	6.75	−0.75	6.00	8	4	5	11	72	80	155
Germany (FRG)	8.85	1.64	10.49	7	26	11	8	47	81	157
Ireland	0.97	0.06	1.03	9	41	0	13	37	115	228
Netherlands	6.33	−1.59	4.74	8	12	54	2	25	83	165
UK	6.60	1.11	7.71	8	7	8	2	75	103	205
Poland	39.51	−0.24	39.27	14	53	9	14	11	119	238
USSR	77.71	1.91	78.62	22	27	6	8	37	110	213
Kenya	0.30	0.00	0.30	11	0	0	15	74	14	27
Madagascar	0.17	0.00	0.17	17	5	0	12	66	13	25
Rwanda	0.22	0.00	0.22	16	0	0	10	74	34	66
Argentina	1.84	0.05	1.89	6	0	0	4	90	63	115
Mexico	0.98	0.00	0.99	6	9	0	0	10	84	12
Peru	1.59	0.00	1.59	13	0	0	10	77	70	140
Algeria	0.54	0.20	0.74	12	3	0	9	76	30	57
Turkey	2.96	−0.01	2.95	9	2	0	10	79	51	99
Bangladesh	0.94	0.00	0.94	11	0	0	10	79	9	16
China	46.70[a]	−0.13	46.53	9	15	15	5	56	26	50
India	9.35	−0.02	9.33	15	0	0	16	69	10	18
North Korea	1.54	0.00	1.54	7	5	5	10	73	63	122

[a] Adjusted to reflect the most recent FAO estimate.

Source: FAO, *Food balance sheets 1979–1981 average* (Rome, 1984).

nearly parallel. Since 1960, in most areas, per capita potato consumption has grown somewhat more rapidly than per capita production as better technology has lowered post-harvest losses and as increasing yields have reduced the share of the crop that must be reserved for seed.

In contrast to the immense international trade in the major cereal grains, only a small portion of the world's potatoes—about 2 percent—are exported on international markets. Potato exports are insignificant in all but a few countries because their bulkiness and perishability make potatoes costly and risky to haul over long distances. Also, quarantine regulations restrict international trade in potatoes in some instances.

Seeding rates significantly affect the amount of potatoes available for human consumption. Most potato growers must plant 1 or 2 tons of seed tubers per hectare, which is commonly 10 to 15 percent of the harvest, but can be over 20 percent.

Potato consumption levels vary greatly among countries and regions (Fig. 3). Although potato consumption in Europe has been falling, it still exceeds 100 kilograms per year per head in the UK, Poland, the USSR, and East Germany. And although consumption is expanding in the Third World, most people in developing countries still eat less than 20 kilograms of potatoes a year. Within the developing world, average potato consumption levels are lowest in the hot tropics and highest in countries that have significant temperate or highland production zones. Potato consumption is rising particularly rapidly in new areas—areas where the demand for food is growing fast and technological change has lowered unit production costs, making potato growing profitable for the first time.

Productivity and value of the potato crop

An important reason why European farmers expanded potato cultivation two centuries ago and why farmers in developing countries are doing so today is that, with appropriate technology, the potato crop is highly productive.

As a means of comparing the productivity of crops, fresh-weight yield has little meaning because of large differences in

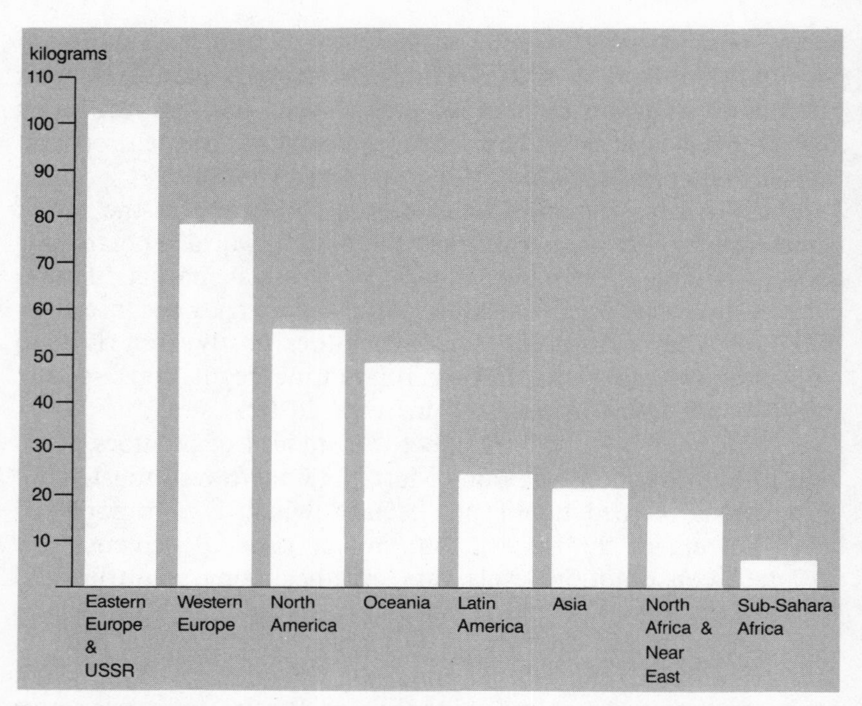

Figure 3. Per capita potato consumption, by region, 1979/81. ***Source:*** **FAO,** ***Food balance sheets 1979*** **(Rome, 1984).**

water content. Root crops, for example, have up to 80 percent moisture at harvest, whereas cereals have around 15 percent. Dry matter production per hectare is a more meaningful yardstick for comparing crops, regardless of their use for food, feed, starch, or alcohol. Measures of edible energy and protein production per hectare are more appropriate indicators of the nutritional yield of crops consumed by humans. The monetary value of production per hectare is a useful criterion for comparing the economic value and income-generating capacity of different crops.

Potato yields are now, on average, 30 percent higher in the developed countries and 80 percent higher in the developing countries than they were in the 1950s. This suggests that in the post-war period, potato technology has improved more substantially in developing countries than in developed countries. Despite these trends, potato yields in developing countries are still less

TABLE 5
The Ten Food Crops with the Highest Production Value per Hectare in Developing Countries

Crop	Dry matter (t/ha)	Edible energy (million kcal/ha)	Edible protein (kg/ha)	Production value (US$/ha)
Tomatoes	1.1	3.1	157	3,159
Cabbages	1.3	3.2	175	3,026
Potatoes	2.3	7.1	196	1,633
Yams	2.6	8.4	175	1,581
Sweet potatoes	4.0	12.6	187	1,210
Cassava	3.4	7.3	32	595
Cocoyams	1.2	3.7	72	554
Rice, paddy	2.6	7.1	130	493
Bananas	1.5	3.9	36	492
Groundnuts in shell	0.9	4.1	190	297

Source: FAO, *Production yearbook 1983* (Rome, 1984); USDA, *Composition of foods* (Washington, D.C., 1975); and FAO estimates of farmgate prices (unpublished). Production estimates are 1981/83 averages; price estimates are for 1977.

than half the levels obtained in North America (as shown previously in Table 3).

Yet it is erroneous to conclude that gaps in yields among countries or regions could be easily closed by transferring technology from the high-yielding areas. One reason is that the *economically optimal* yield is also lower in most developing countries than in Europe and North America. Furthermore, because of differences in environmental conditions, levels of development, and prices, the scope for direct transfer of technology from developed to developing areas is also limited.

In terms of dry matter production per hectare, potatoes are among the most productive crops grown in the developing countries (Table 5). Because of their relatively short growing period, potatoes rank even higher in terms of dry matter production per day (Table 6). In terms of the nutrients available for human consumption, potatoes are also highly productive in relation to other crops grown in developing countries. Potatoes

TABLE 6
The Ten Food Crops with the Highest Production Value per Hectare per Day

Crop	Growth duration (days)	Dry matter (kg/ha/day)	Edible energy (000 kcal/ ha/day)	Edible protein (kg/ha/day)	Production value (US$/ ha/day)
Cabbages	110	12	29	1.6	27.50
Tomatoes	125	8	25	1.3	25.30
Potatoes	130	18	54	1.5	12.60
Yams	180	14	47	1.0	8.80
Sweet potatoes	180	22	70	1.0	6.70
Rice, paddy	145	18	49	0.9	3.40
Groundnuts in shell	115	8	36	1.7	2.60
Wheat	115	14	40	1.6	2.30
Lentils	105	6	23	1.6	2.30
Cassava	272	13	27	0.1	2.20

Source: FAO, *Production yearbook 1983* (Rome, 1984); USDA, *Composition of foods* (Washington, D.C., 1975); and FAO, *Report of the agroecological zones project* (Rome, 1978). Production estimates are 1981/83 averages; price estimates are for 1977.

are conventionally considered to be an inexpensive source of energy, but in fact, they are relatively expensive in most developing areas. Because of the potato crop's combination of high yields, short vegetative cycle, and high price, the value of potato production per hectare exceeds that of most other food crops grown in developing countries, ranking third after cabbage and tomatoes and far ahead of most field crops.

Production of dry matter, edible energy, protein, and monetary value can also be used to measure the total production of food crops (Table 7). In terms of the first three, the potato ranks ninth. But when all commodities are valued at their farmgate prices, the potato comes out fourth on the list of the developing countries' most valuable food crops.

Bibliographic notes

This chapter is based on two main sources of information: publications on the origin and worldwide spread of the potato crop and statistical reports issued by FAO. The most comprehensive historical study of the potato is Salaman (1986). More

TABLE 7
Production of the Ten Highest Value Food Crops in Developing Countries, 1981/83

Crop	Dry matter (million tons)	Edible energy (trillion kcal)	Edible protein (million tons)	Production value (billion US$)
Rice, paddy	355	982	18.1	68.7
Wheat	157	454	18.1	26.6
Maize	134	476	12.2	18.5
Potatoes	17	51	1.4	11.7
Sweet potatoes	31	99	1.5	9.5
Cassava	51	109	0.5	8.9
Soybeans	28	127	10.8	6.5
Sorghum	42	149	4.9	5.8
Groundnuts in shell	17	74	3.4	5.3
Bananas	13	34	0.3	4.2

Source: FAO, *Production yearbook 1983* (Rome, 1984); USDA, *Composition of foods* (Washington, D.C., 1975). FAO estimates of farmgate prices (unpublished). Production estimates are 1981/83 averages; price estimates are for 1977.

concise treatments of the origin and spread of the potato crop are in Hawkes (1978a, b) and Burton (1966). These publications focus on the Americas and Europe and touch only briefly on potato cultivation in other regions. Mokyr (1983) presents a detailed analysis of the Irish famine. The most thorough discussion of the spread of the potato to other parts of the world is Laufer (1938). The section of this chapter covering origin and spread of the potato crop draws heavily on Hawkes (1978a), Burton (1966), and Laufer (1938). Information on India is from Pushkarnath (1976) and Srivastava (1980).

The remaining sections of the chapter are based on data presented in FAO production yearbooks and food balance sheets. Estimates of the monetary value of food crop production are based on production figures in the production yearbooks and unpublished estimates of weighted average farmgate prices for all developing market economies made available by FAO's Basic Data Unit. I am particularly indebted to K. Becker and his colleagues at FAO for making these available. With respect to the FAO data, it is important to note that estimates of potato production in China were substantially revised in 1978 and again in 1983. All figures appearing in this chapter are based on the FAO's most recent estimates. The FAO's recent major revision of estimates of root crop production in China is based on detailed work at the International Food Policy Research Institute, reported in Stone (1984).

More comprehensive statistics and discussions of world trends in potato production and use are contained in Horton and Fano (1985) and van der Zaag and Horton (1983). Nick Young, at the Center for European Agricultural Studies, Wye College, England, has published a number of useful studies of the potato industry in Europe. Young (1981) provides a summary of several of these.

2
The potato crop and its physical environment

To overcome the principal constraints to potato production and use in developing countries, one must first understand the plant and the physical and socioeconomic environment in which it is grown. This chapter describes the potato plant and the physical factors that influence its growth and yield. Socioeconomic factors affecting potato production and use are discussed in the next chapter.

The potato plant

Botanical aspects

The potato is one of about 2,000 species in the family Solanaceae, which includes such plants as tobacco, tomato, eggplant, and pepper. There are 8 cultivated species of the tuber-bearing solanums and about 200 wild species. All these relatives of the potato are of New World origin.

The word "potato" is derived from *batata,* the Caribbean Arawak name for sweet potato. The Spanish conquistadores encountered the sweet potato first and subsequently gave the same name to other tuberous plants in the Americas. From *batata* evolved the Spanish *patata* and the English "potato."

Confusion of the potato with other root crops continues in many parts of the world. In England in the sixteenth century, both *Ipomoea batatas* (sweet potato) and *Solanum tuberosum* (potato) were indiscriminately called "potatoes," and this custom spread to other English-speaking parts of the world. Until recently, some U.S. statistical publications reported figures for both crops under a single heading labeled "potatoes." In China, potatoes are known by more than 20 names, many of which are also used for sweet potatoes, yams, and other root crops. Until 1983, China's agricultural statistics also placed potatoes and sweet potatoes under a single heading. (To add to the confusion, the Chinese reported root crop production in "cereal equivalents" under the generic heading of "grains"!)

Grouping potatoes with other crops can be very misleading, not only because the plants are botanically distinct but also because the patterns and trends in potato production and use often differ sharply from those for other root crops. For example, in many areas where consumption of sweet potatoes or cassava is falling, potato consumption is rising rapidly.

The potato may be classified as a dicotyledonous annual, though it can persist in the field vegetatively (as tubers) from one season to the next. Some farmers in such widely dispersed locations as Ireland, Newfoundland, Peru, Guatemala, the Himalayas, and Rwanda leave some tubers in the ground to serve as seed for the next crop. In most areas, however, farmers eliminate volunteer plants growing from unharvested potato tubers because they may harbor potato pests or serve as sources of disease infection.

The potato plant has shallow roots—they seldom extend deeper than 40 to 50 centimeters (Fig. 4). If, however, no obstructive layers or sharp transitions from one soil type to another occur in the soil profile, plants may root as deep as 1 meter.

The tuber is an enlarged portion of an underground stem adapted to storage of photosynthates and reproduction of the plant (Fig. 5). The tubers, which originate from the tips of underground stems called stolons, contain all the characteristics of normal stems, including dormant buds (the "eyes") formed at the base of a leaf (rudimentary in this case) with detectable leaf scars (the "eyebrows"). Lenticels or stem pores through which air penetrates to the stem interior are found on tubers. The eyes occur in a spiral pattern on the tuber, with few near the attachment of the tuber to the stolon and most toward the end of the tuber, known as the apical end. The apical buds

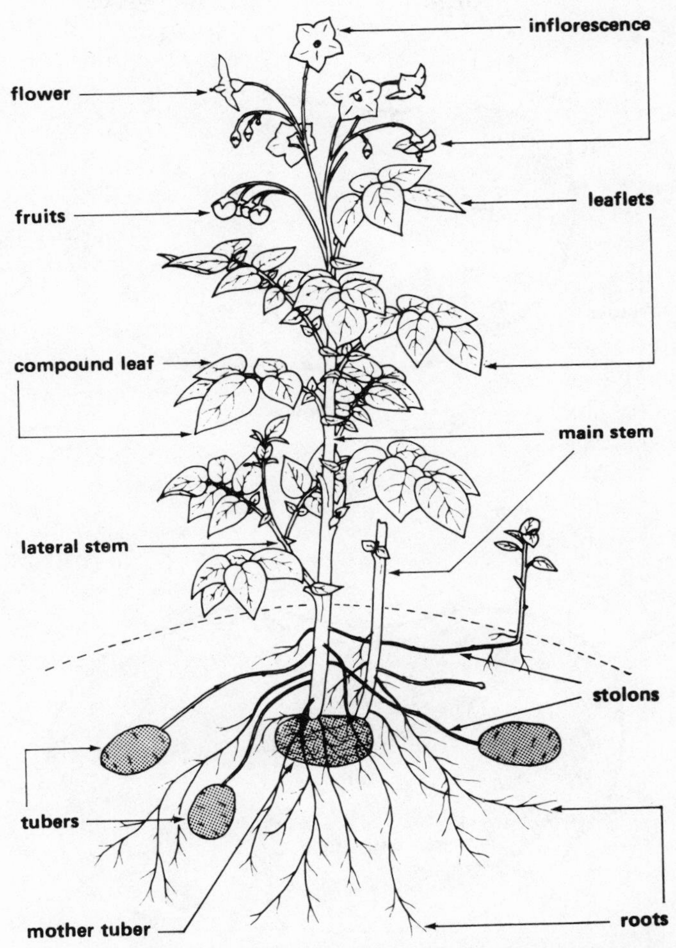

Figure 4. The potato plant: foliage, root system, and tubers. *Source:* Zosimo Huamán, *Systematic botany and morphology of the potato,* Technical Information Bulletin No. 6. 2nd ed. (Lima: International Potato Center, 1986).

(eyes) possess dominance and will normally sprout first. When the apical buds are removed, or die, other buds are stimulated to sprout. As will be seen below, apical dominance and tuber physiology in general have an important bearing on production and use of "seed" tubers.

The outer layer of the tuber's cells is known as the epidermis. Immediately below the epidermis is the periderm consisting of

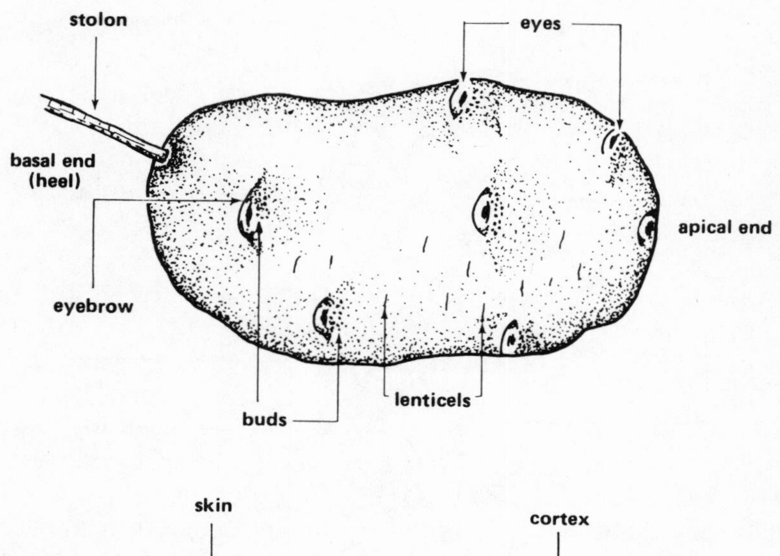

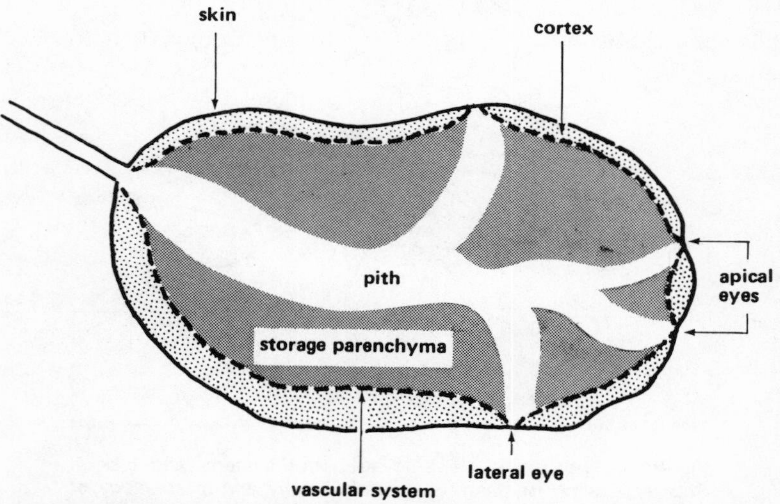

Figure 5. Major parts of the potato tuber. *Source:* Zosimo Huamán, *Systematic botany and morphology of the potato,* Technical Information Bulletin No. 6. 2nd ed. (Lima: International Potato Center, 1986).

several layers of corky cells. The epidermis and periderm together constitute the tuber's "skin." The skin of a mature tuber is practically impermeable to chemicals, gases, and liquids, provides good protection against microorganisms, and resists water loss. But if potatoes are grown in waterlogged soil, the lenticels of the tubers open wide, permitting easy entry of harmful microorganisms. Tubers harvested without a well-developed skin are easily damaged and will lose moisture rapidly during storage; microorganisms can also easily enter such "immature" tubers. If a tuber is injured or cut, the tissue may also suffer from infection and excess moisture loss until the wound heals—a process that can take 3 to 5 days under favorable conditions (an environment with sufficient oxygen, temperatures between 10 and 20 degrees C, and around 90 percent relative humidity).

The remainder of the tuber from the vascular ring inward, designated as the medullary area, is divided into outer and inner medulla and constitutes the fleshy part of the tuber. The outer medulla includes the watery and more translucent part. The inner medulla extends toward each eye, forming a continuous tissue that connects all the eyes of the tuber.

About 2 to 4 weeks after the stems first emerge from the soil, young tubers begin to grow at the tips of the stolons. Growth of the young tuber is the result of cell division and elongation and storage of translocated photosynthates within the cells. In order for tubers to enlarge—a process called bulking—the amount of photosynthate available for translocation and storage must exceed that needed by other parts of the plant for growth and metabolism. This process is affected by many environmental factors.

Although tuber formation, or tuberization, is not dependent upon flowering, the two are sometimes associated because conditions that are unfavorable for flowering, such as hot and dry weather, also retard tuber formation. Conditions that favor flowering usually favor tuberization as well.

Vegetative reproduction

The potato crop can be reproduced sexually by planting tiny true seeds, which form in the small, tomato-like fruits. However, with few exceptions, the world's potato growers plant tubers. This type of vegetative reproduction is one of the distinguishing features of the potato crop. It strongly influences not only how potatoes are grown but also where, by whom, and for what uses. In most of the developing world, the scarcity and high cost of good-quality seed tubers is a serious constraint to potato production.

Seed tuber physiology. The yield of a potato crop is affected by the number of main stems per hectare, which in turn depends on the number of sprouts that form from the eyes of seed tubers. The number of eyes on a tuber varies, depending on the variety, tuber size, and the environmental conditions. For a given variety, the number of eyes is roughly proportional to the surface area of a tuber. Hence, small tubers have more eyes per unit weight than large ones (small spheres have more surface area per unit weight than large spheres of the same density). For this reason, most potato farmers prefer to plant small seed tubers.

The number of eyes that sprout is related to the physiological stage of the tuber when it is planted. After being harvested, tubers pass through four stages: dormancy, apical dominance, maturity, and senility. Dormant tubers produce no sprouts; during apical dominance, only the apical sprout grows; in maturity, many strong, vigorous sprouts grow; and in senility, although some sprouts form, they tend to be thin and to produce weak, low-yielding plants.

The length of a tuber's dormancy period depends on the variety, the environmental conditions under which it was grown, its maturity at harvest, and how and for how long it was stored. Varieties differ markedly in length of dormancy, and this characteristic often influences the varietal preferences of farmers.

Potatoes grown under high temperatures, especially at the end of the growing period, and under short-day conditions (as in the plains of India and Pakistan) also tend to have a shorter dormancy period than those grown in areas with long days and cool temperatures (as in northern Europe).

Storage conditions, particularly temperature, have significant influences on dormancy and sprout growth. Tubers stored under high temperatures, high relative humidity, and in darkness have short dormant periods. Tubers that have been attacked by microorganisms or insects or that have been damaged, by cutting for example, also have a shorter dormant period than healthy, undamaged tubers.

If a tuber starts sprout growth during the apical dominance stage, only one of the buds at the apex will develop a sprout.

Removal of this top sprout allows other buds to develop sprouts. Thus farmers often desprout seed tubers before planting to break apical dominance and ensure uniform emergence of several main stems. If a seed tuber is stored at low temperatures and sprout growth starts in the maturity stage, many eyes will develop sprouts.

Storage temperature and exposure of tubers to light strongly influence sprout growth and the resulting vigor of seed tubers. Sprouts occur on tubers stored at temperatures higher than about 4 degrees C. Excessive sprout growth causes dehydration of the tubers and reduces the vigor of the crop grown from them. Light retards sprout growth, and potatoes that are stored in light develop green sprouts that are much shorter and sturdier than sprouts on tubers stored in the dark. Consequently, storing seed tubers in light can compensate for some of the negative effects of high storage temperatures. This aspect of tuber physiology has practical implications for seed storage in developing countries. That is, the storage life of potatoes held without refrigeration can be prolonged by exposing them to light. Light storage cannot, however, be used for consumer potatoes, since greening makes them taste bitter.

Tuber size affects sprout growth in two ways. Large tubers produce more sprouts and main stems than small ones because their development is positively correlated with the tuber's surface area. Sprouts also grow faster on large tubers because they have a greater food reserve available for each sprout than small tubers do. The smaller the seed tuber and the larger the number of sprouts developing on it, the greater the competition between sprouts and the slower their growth rate. However, per unit weight, small seed produces higher yields than large seed of equal health and physiological condition.

The period between planting and crop emergence is the most delicate stage of the potato crop's growth. Therefore, yields are strongly influenced by the physiological stage of seed tubers at planting time. Tubers should be sprouted before planting, so that emergence and development of several strong stems will take place soon after.

The environmental conditions under which seed tubers are produced and stored have a strong influence on the vigor and growth cycle of the resultant crop. Seed grown in cool weather and stored at low temperatures and high humidity may behave as relatively young seed even after 8 or 9 months of storage. Young seed tends to be more vigorous than old seed and produces a later-maturing crop with a higher potential yield.

Disease transmission. Many yield-reducing diseases are transmitted by seed tubers. Some, such as late blight, are easily spread during the growing season. Others, such as bacterial wilt, can also remain in the soil for several years.

The most important diseases transmitted by seed tubers are virus diseases. Preventing virus spread is the chief objective of seed potato certification. Viruses depress potato yields by reducing the amount of green foliage and photosynthesis. The magnitude of yield losses from virus diseases is related to the type of virus, the virus tolerance of the variety planted, and the growing conditions. When all plants in a field are infected with an aggressive virus like potato leaf roll virus, yields may be cut by more than 50 percent. Andean varieties (*S. tuberosum andigena*), which produce abundant foliage, are normally less affected by virus infection than European and North American varieties (*S. tuberosum tuberosum*), which have less foliage. Growing conditions—soil, weather, fertilization—are also factors because, in a well-developed crop, the foliage of healthy neighboring plants will partially compensate for the decreased foliage of the diseased plants.

Early and late crops

Potato varieties differ considerably in the length of time needed to reach maturity. Most varieties of the subspecies *andigena* have a long growing period (4 to 6 months), whereas varieties of the subspecies *tuberosum* mature more quickly. Within the *tuberosum* group, some varieties mature much earlier than others.

The variety farmers choose must fit the length of the cropping season in their locations and their cropping alternatives. In areas where the growing season is short, early potatoes are usually grown. Where the season is longer, a later-maturing variety is usually preferred. But where a long season permits double or triple cropping, an early-maturing, short-season potato variety

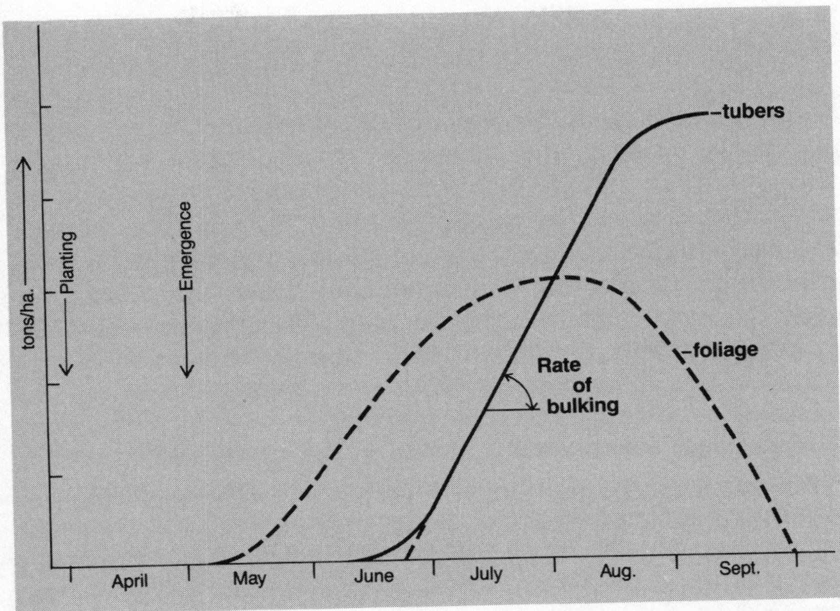

Figure 6. Growth cycle of a potato crop. Adapted from: H. P. Beukema and D. E. van der Zaag, *Potato improvement* (Wageningen: International Agricultural Centre, 1979).

may be desirable. A disadvantage of late varieties, which needs to be kept in mind, is that their prolonged exposure to drought, pests, and other hazards may limit yields or lower the quality of harvested tubers.

Three periods can be distinguished in the potato's growth cycle: preemergence/emergence, foliage growth, and tuber growth (Fig. 6). Foliage growth and tuber growth may overlap for a considerable time, especially in late-maturing varieties. Stems grow from the sprouts of the seed tuber. After stems emerge from the soil, foliage and roots develop simultaneously and their growth is correlated. Tubers generally start growing slowly about 2 to 4 weeks after emergence and continue growing at a fairly steady rate. Early-maturing varieties have early emergence and moderate growth of foliage. They produce a relatively high yield in a short period. Late varieties, in contrast, have later emergence and more abundant foliage. If they are harvested early in the growing season, late-maturing varieties produce a rather low yield, but when harvested later in the season they outyield earlier varieties.

The length of a potato variety's growth cycle is influenced by environmental conditions, so that a variety that is late under one set of growing conditions can be early under another. Farmers also have some latitude in prolonging or shortening the periods in a variety's growth cycle. For example, by applying nitrogen and irrigating heavily, they can extend the vegetative period and raise yields. By planting well-sprouted seed they can shorten the cycle. Farmers often adjust their agronomic practices in order to make available varieties fit their cropping system.

The physical environment

Environmental conditions influence whether farmers will grow potatoes and, if so, how. They also affect yield levels and how the harvest is used. Three groups of environmental conditions are of paramount importance: weather and climate; soils, fertilization, and relief; and pests and diseases.

Weather and climate

Daylength and temperature. Potatoes are grown as an economic crop under a wide range of daylength regimes—from 12 hours of sunlight in the Andes and equatorial zones of Africa and Asia, to over 16 hours of sunlight in Alaska at 60 degrees north latitude and in Punta Arenas, Chile, at 53 degrees south latitude. Daylength and temperature can influence the growth habit of the potato in unexpected ways. For example, in varieties of the *tuberosum* subspecies, short days and moderate temperatures, particularly low night temperatures, stimulate tuber initiation; varieties of the *andigena* subspecies, however, mature late under short days. Soil temperatures of 15 to 18 degrees C appear to be the most favorable for common potato varieties.

The daily bulking rate is related to the hours of daylight; hence, bulking tends to be more rapid in long-day temperate zones than in the tropics. Differences in final yield are offset by a longer growing season in some tropical areas, but in others both the daylength and the growing period for potatoes are shorter than in temperate zones. For these reasons, as well as to minimize exposure to pests and climatic hazards, early-maturing varieties are usually desirable in the tropics.

Two aspects of temperature are particularly significant for potato production. High *daytime* temperature is usually correlated with a high rate of respiration and transpiration, which may cause plant moisture stress even when the soil moisture content is high. Furthermore, the rate of net photosynthesis decreases when temperatures are higher than about 25 degrees C.

The second and more critical variable is minimum *nighttime* temperature. As a rule, tubers will not begin to form if nighttime temperatures remain above 20 degrees C. High night temperatures increase plant respiration, depleting carbohydrate reserves and slowing tuber growth.

Sprout growth is also influenced by soil temperature. Soil temperatures below about 12 degrees or above 28 degrees C impair sprout development.

Frost and hail can damage potato plants and cut yields drastically. Farmers high in the Andes grow "bitter" potatoes because these potatoes tolerate frost better than other crops. Potatoes grown elsewhere in the world are vulnerable to frost, though they are otherwise well adapted to cool climates.

Moisture requirements and irrigation. The potato plant is more sensitive to drought than most other crops. Only a small part of the water taken up by the plant is used directly in photosynthesis. The primary roles of water are to cool the plant by evapotranspiration and to provide a medium for transporting organic compounds and minerals within the plant.

High relative humidity stimulates root formation on potato sprouts. Hence, to encourage rapid emergence, a liberal amount of water should be available in the soil at planting time. After emergence, the availability of water to the potato crop continues to affect yield because of the potato's shallow root system and the negative effect of water stress on photosynthesis. In general, a dry spell that occurs in the later stages of growth reduces yield more than one during earlier stages. In warm periods potatoes require more frequent irrigation than other crops. Farmers who are inexperienced with potatoes may be unaware of this char-

In order to minimize frost damage to early seeded potatoes in South Korea, farmers plant under plastic.

acteristic and therefore underirrigate the crop. During cool periods, potatoes are better able to tolerate brief droughts.

Excessive soil moisture, although less common than moisture shortage, may also reduce potato yields. The limited oxygen content of very wet soil damages roots, and new tubers in saturated soil may rot because their lenticels open, permitting bacteria to enter.

Irregular water supply induces growth cracks, "hollow heart," irregular growth, and other malformations that may decrease the market value of tubers.

Soils

The potato crop develops best on deep, friable soils that have good water retention. Because the potato has a relatively weak root system, impermeable layers in the soil limit rooting depth, which, in turn, restricts availability of water to the plant in dry periods. As a result, soil compaction can greatly reduce potato yields. Another penalty of soil that has compacted or dense layers is that it readily becomes saturated after heavy rainfall or irrigation, killing the roots and causing the tubers to rot.

Fertilization

Because potatoes respond extremely well to both farmyard manure and mineral fertilizers, most growers apply fertilizers. Although recommendations on the amount and ratio of fertilizer nutrients to apply to potatoes must be based on local experience and trials in farmers' fields, a few generalizations can be made. Nitrogen stimulates growth of foliage and delays tuber formation. Hence, a well-fertilized crop will mature later in the season and yield more than one with less nitrogen. If harvested early, potatoes from a heavily fertilized crop may be damaged and difficult to store. For these reasons, farmers ordinarily use higher rates of nitrogen on the main-season or late crops and lower rates on early crops (which are harvested immature to fetch a high price) and seed crops (for which small tubers are desired).

Application of phosphorus contributes to early tuberization and development of the crop. Added potash often has less effect on yield than added nitrogen or phosphorus, but it may reduce the dry matter content of the tubers. Potatoes are tolerant of soil acidity, but in soils that have a pH below about 4.8, yields can be severely depressed by calcium deficiency.

Potatoes benefit from the nutrients in farmyard manure but even more from the added organic matter. Humus improves the structure of heavy soils and enlarges the moisture retention capacity of light soils. In addition, supplying farmyard manure enhances the effect of nitrogen provided in chemical fertilizer.

Relief

In crop production, differences in altitude compensate in some ways for differences in latitude. Because average temperatures are inversely correlated with elevation, European potato varieties sometimes do well in highland tropical areas. As farmers begin growing earlier varieties adapted to warmer growing conditions and using cultural practices that moderate the adverse effect of warm weather on growth of the potato crop, potato production is spreading in tropical areas.

As a rule, soil scientists and extension workers recommend that field crops, including potatoes, be planted on level land rather than on hillsides to minimize erosion and take advantage of fertile valley-bottom soils. Where slopes must be cultivated, planting on horizontal "contour lines" is recommended. To the dismay of the experts, farmers in many tropical highland areas plant their crops on the hillsides and let livestock graze the flat bottomlands. To make matters worse from the point of view of soil scientists, farmers often orient their rows up and down the slope.

Although in the long run, cultivation of sloping land fosters soil erosion, farmers often have good reasons for growing potatoes on hillsides. Due to increasing land scarcity and population pressure, farmers are forced to intensify crop production in mountainous areas. When night frost occurs, cold air moves down the slopes and concentrates on the valley floors. Therefore, by planting crops on the slopes, farmers in these areas can lower the risk of frost damage. And although cultivating the slopes is bound to cause erosion and should be discouraged, erosion may in fact be minimized by planting up and down or diagonally to the slope, rather than laterally along the contour. Farmers have learned that rows planted on the contour impede rapid drainage and may be completely washed away when torrential rainstorms occur.

Pests and diseases

The potato crop is susceptible to over 300 pests and diseases, but it is neither practical nor necessary to control them all because most have a negligible effect on production. Indiscriminate "blanket" control can do more harm than good by destroying beneficial organisms and upsetting delicate ecosystems. As a result, farmers try to control only those pests and diseases that they feel cause extensive crop losses, survive a long time in the soil, or attack other crops in addition to potatoes.

Diseases can spread through seed, soil, implements, insect carriers (vectors), and other means. Because of the importance of seed tubers as a source of infection, quarantine systems and national seed certification programs are often established to control the spread of diseases.

Virus diseases. Viruses are found in all potato-producing regions. Once infected, potato plants cannot be cured. Consequently, methods for controlling viruses center on the use of virus-resistant varieties and detection and elimination of infected plants and their tubers from seed fields and stocks. Some of the costliest virus diseases, such as leaf roll, are commonly transmitted by insects. Generally, tubers produced in cool, rainy and windy areas (such as Scotland, southern Chile, and tropical highland zones) have little virus infection because populations of virus-transmitting insects, particularly aphids, are low in these areas.

Bacterial diseases. Some bacterial diseases are both seed- and soil-borne. Generally the most destructive one in developing areas is bacterial wilt, caused by *Pseudomonas solanacearum.* Bacterial wilt is endemic in much of the lowland humid tropics and is a major barrier to potato cultivation. Many farmers try to ward off infection of their fields by purchasing seed only from areas free of bacterial wilt. In areas that are already infected, rotations can be used to help avoid buildup of bacterial diseases. When severe outbreaks occur, government agencies sometimes impose quarantines to keep seed tubers from moving to noninfected areas. Nevertheless, because the pathogen attacks hundreds of plant species, many of them also cultivated, quarantine is rarely very effective.

Fungal diseases. Among the fungal diseases, late blight, caused by *Phytophthora infestans,* is the most widespread and damaging. It is, in fact, the most important disease of the potato, worldwide. Infected crops can seldom be cured. Control depends on use of resistant varieties or chemical sprays.

Since the nineteenth century, it has been known that treating the foliage of potato plants with copper compounds helps protect them from late blight. Dozens of protective fungicides are now available for foliar application against late blight.

To prevent the buildup of lesions that serve as sources of infection, the fungicide must be present on the foliage at the time of inoculation. Farmers often begin spraying early in the

Fungicides are heavily applied by potato growers in Costa Rica and many other countries. Research can reduce the cost of inputs farmers use.

growing season before the first attack is expected. The frequency of spraying depends on local conditions. Intermittent rainfall, sunny days, and moderate to warm temperatures are conducive to blight attacks. Where blight occurs, and where fungicides are available, farmers usually spray every 3 to 20 days, depending on the probability and expected severity of the attack.

Few European or North American potato varieties are resistant to late blight, but several varieties bred in developing countries are. The most notable progress in selecting for late blight resistance has been made in Mexico, where several new resistant varieties have been identified and distributed to other countries since the 1950s. Low-income farmers in more than a dozen countries, including Costa Rica, Rwanda, Nepal, and the Philippines, now grow these varieties. These varieties permit farmers to grow potatoes in rainy areas or seasons in which potato cultivation was never before possible. Owing to the high cost and toxicity of fungicides, breeding for blight resistance is an important strategy for control of this disease.

Insects and nematodes. In many potato growing areas, as fallow periods have been eliminated and rotations have intensified, insect populations have increased. Many farmers have gotten onto a treadmill that requires them to apply ever-increasing quantities of pesticides just to maintain the same level of control. Often they find that even at higher rates of application and expense, insecticides provide less control than they did a few years ago.

The potato tuber moth, *Phthorimaea operculella,* is considered the leading pest of potatoes in developing countries. Several years ago, it was a problem mainly in warm environments, but several infestations have now been reported in cooler areas, like highland Colombia, Kenya, Nepal, and Peru. It is likely that it has now reached every developing country where potatoes are grown. The insect has high reproductive potential, and its resistance to insecticides appears to be rising. Most potato varieties are highly vulnerable in the field and in storage.

Insects damage crops both directly and indirectly. Where insect infestation levels are high, direct mechanical damage to foliage and tubers may seriously lower yields and quality. Some insects also harm crops indirectly by transmitting diseases. Aphids, for example, spread viruses, which cut yields of future potato crops because viruses are transmitted through seed tubers.

Nematodes, microscopic wormlike organisms that bore into tubers, cause both primary and secondary damage. They decrease yields and tuber quality in infested crops and are spread through seed tubers to subsequent crops. Farmers who grow potatoes for market concentrate on preventing direct damage by insects, but farmers growing potatoes for seed need to avert both direct and indirect damage.

Two major nematodes of economic importance for potatoes in developing countries are cyst nematode (*Globodera* spp.), particularly in the highland tropics, and root-knot nematode (*Meloidogyne* spp.), mainly in the lowland tropics. The cyst nematode also occurs in Europe and North America. Although the general distribution of the major nematode species is known, there are no systematic data on incidence and crop loss. Nematode control is also made difficult by the wide range of pathogenic variability and the fact that nematodes attack many other plants and live several years in the soil.

Plant-host resistance is potentially one of the most effective and economic means for controlling nematodes. Breeding for cyst nematode resistance began in Europe after discovery of resistance in a few native Andean varieties and related weedy species. However, resistance incorporated into varieties in Europe is ineffective in the Andean region. Breeding programs exist in Europe, the United Kingdom, the USSR, the United States, Colombia, Ecuador, India, and Peru. Panama is screening varieties for nematode resistance.

Breeding new potato varieties

The genetic diversity of cultivated and wild potatoes is greater than that of any other major world food crop. In addition to *S. tuberosum,* there are 7 other cultivated species and over 200 wild species of potato. In contrast, there is only one cultivated and no known wild species of maize. Rice and wheat have no more than eight species. Considerable diversity is found within the root crops like cassava, sweet potato, and yams, but in none of these does the genetic diversity approach that of the potato.

Most cultivated potatoes in the world are tetraploids with 48 chromosomes ($2n = 48$). In the Andes, however, cultivated diploids ($2n = 24$), triploids ($2n = 36$), and pentaploids ($2n = 60$) are also found. Wild species with 24, 36, 48, 60, and 72 chromosomes occur in Central and South America.

Due to their broad genetic background, cultivated potatoes are extremely heterozygous: the offspring, or progeny, resulting from a cross of two parents, even within the same variety, are usually highly varied. For this reason, varietal purity can only be maintained by asexual (or vegetative) reproduction using tubers or other parts of stems as planting material.

One practical consequence of the potato's heterozygosity is that the probability of selecting an offspring that is superior to either of its parents is extremely low. Potato breeding programs usually need to make several hundred crosses and evaluate hundreds of thousands of seedlings (grown from seeds produced by the crosses) and clones (derived asexually from selected seedlings) over several years—often 10 or more—before releasing a single new variety.

Only a few potatoes, representing a tiny fraction of the genetic diversity within *S. tuberosum,* were taken to Europe by the Spanish conquistadores. Part of the initial genetic variability was lost because some Andean varieties did not tuberize under Europe's long-day conditions. Late blight epidemics in the nineteenth century eliminated other varieties, further reducing the genetic base of European potatoes.

The narrow genetic base of the European varieties has stunted many potato breeding programs. Many breeders have consequently turned to the Andean potato germplasm—wild species and native varieties—as sources of resistance to major pests and diseases and as sources of adaptation to environmental conditions like frost, heat, and drought. New approaches have been developed for breeding at the diploid level and returning later to the tetraploid level.

The application of genetic engineering (recombinant DNA) technology has been successfully applied to potatoes. The gene vector systems are clearly functional. However, to date, few purified genes that could improve potato quality are available. The major advantage of genetic engineering is that it allows one or more desirable traits—like virus resistance—to be added to an existing variety, while maintaining the other traits such as agronomic adaptation and culinary quality.

"Population breeding," based on large-scale crossing and recurrent selection with progeny testing, is used by the International Potato Center to develop improved breeding material with wide genetic diversity to assure high yields and stability of performance, as well as increased frequency of genes controlling resistances, adaptation, yield, and desirable quality factors. Advanced lines and selected clones are made available to scientists in national potato programs, who can then make further crosses or select clones that best meet their needs.

Several national programs have released new varieties recently: Molinera, Caxamarca, Perricholi in Peru; Bastides in Ecuador; Kinigi and Nseko in Rwanda; Naataange in Senegal; Muziranzara in Burundi; Dalat 004, 006, and 012 in Vietnam; Domoni in Fiji; and Sita and Krushi in Sri Lanka. The breeding and selection work on which these varieties are based was done in Argentina, Mexico, Peru, and other country programs as well as at CIP.

Yield and value of the harvest

Crop yields are customarily measured in physical weight per unit of land area. However, potato farmers, traders, processors, and consumers are less concerned with total production than with the size of tubers and their quality, as reflected in skin and flesh color, dry matter content, taste, texture, and signs of damage caused by pests, diseases, and handling.

Total yield is determined by the length of the growing season and the average tuber production per day. Maximum yields require a high level of daily production over a long period. In some areas, farmers purposely sacrifice yields by killing the foliage and harvesting potatoes early in order to clear the field for another crop (as in South Korea), to get a high market price (as in Bangladesh), or to obtain food when other supplies are scarce (as in Burundi). Thus average daily production, the difference between assimilation and respiration, is strongly influenced by farmers' cultural practices. To obtain high yields, farmers need to plant the right variety for their location, use healthy seed that is in good physiological condition, and pay close attention to soil moisture, fertilization, and pest management.

The quantity and size of tubers produced per hectare is determined by the number of tubers produced per stem and the number of stems per hectare (stem density). The number of tubers produced per stem depends largely on the variety planted and soil and weather conditions. The stem density depends on the variety, number of viable sprouts planted per hectare, sprout damage at planting time, and growing conditions.

Tuber quality is influenced by the variety grown, weather, incidence of pests and diseases, and cultural practices. Decay of tubers and fungal and bacterial problems in storage may stem from waterlogged soils or rainy weather at harvest time. Growth cracks are caused by irregular water supply during the season. Under similar growing conditions and management, different varieties produce tubers differing in size, dry matter, and taste.

The price potatoes command depends not only on their physical attributes but on how they will be used, on local tastes and preferences, and on market conditions. Consumers and processors generally pay premiums for specific varieties, size grades, and qualities of potatoes. Price ranges as great as 5:1 have been observed for different types of potatoes in the Lima market on a given day. The native potatoes selling for the highest price, with a dark blue skin and deep eyes, would be considered unfit for human consumption in many other parts of the world. In many places, consumers prefer large tubers, but in Indonesia, for example, small tubers are generally preferred. Farmers usually want small tubers for seed, because with a ton of small seed tubers they can plant a larger land area than with a ton of large tubers. Thus the price per kilogram of seed tubers tends to be inversely correlated with the average tuber size. Damaged tubers sell for less, but the importance that consumers attach to specific types of damage differs from place to place. Processors have special requirements for potatoes. Starch manufacturers prefer potatoes that have high dry-matter content because they yield more starch, but producers of fried foods prefer potatoes that have low dry-matter content because they absorb less cooking oil.

In selecting varieties and evaluating new production methods, researchers need to look beyond total yield to the quality factors that influence the economic value and acceptability of potatoes. For new varieties to be successful with producers and consumers, the standards breeders use to measure such things as "tuber quality" and "marketable yield" should reflect local uses of potatoes, market conditions, and consumer preferences. The use of standards that do not reflect local conditions is an important reason many new "improved" varieties are ignored by farmers.

Economic implications

The genetic diversity found in potatoes means that through breeding and selection new varieties can be produced that grow well under a wide range of ecological conditions. Varietal improvement has been a key factor in expanding potato production throughout the world. Nevertheless, certain environmental conditions prevent the cultivation of existing potato varieties. The requirements that night temperature be below 20 degrees C for tuberization means that the potato cannot be grown as a food crop in the vast lowland tropics that have high night temperatures.

Generally, the potato can be characterized as a high-input, high-output, high-risk crop. The great responsiveness of yields to inputs—such as high quality seed tubers, fertilizers, pesticides, additional labor, and other forms of energy—motivates farmers to use inputs more heavily on potatoes than on other crops. Because of the relatively high level of potato yields, the short growing period, and the high market value of potato tubers, the potato crop generates larger returns per hectare and per day than most other crops grown in developing countries. The susceptibility of the potato to pests, diseases, moisture stress, and extremes in weather makes its yields more variable than those of many other crops. Yield variation, coupled with price fluctuations and high input costs, makes potato production risky. Due to the high costs and risks, few farmers specialize in potato production; most grow potatoes as one component of farming systems that include several other crops and livestock.

The high moisture content and perishability of potato tubers makes them expensive to transport, store, and process. Marketing costs for potatoes are lower than those for highly perishable fruits and vegetables, but higher than those for cereals and pulses, which have a lower moisture content. Potato marketing is particularly costly in lowland tropical and subtropical areas where potatoes are harvested in spring and market operations take place during the hot summer months. Improvements in storage have stimulated expansion of potato growing in many such areas.

Bulkiness and perishability are the main reasons so few of the world's potatoes are exported or imported. Transportation costs between most countries exceed the differences in production costs. The high risk of spoilage also discourages international trade, particularly in developing areas where temperatures are high; refrigerated storage and shipping facilities are limited; and coordination of truck, rail, and ocean shipping often is difficult.

The variablity of potato yields, perishability of harvested tubers, and high costs of transportation and storage result in price instability. Price fluctuations are greatest in developing countries, but even in developed countries, potato prices fluctuate more than prices of cereals.

The vegetative reproduction of potatoes has several economic implications. Unlike crops grown from seeds, planting material for potatoes is often the single most costly input, accounting for a fourth or more of the production expenditures in most developing areas. If seed tubers are imported, they sometimes may account for over half the total cost. Clearly, less expensive planting material could substantially cut production costs. In many developing areas healthy seed is not available in the appropriate physiological stage (maturity) at the time when farmers would like to plant potatoes. Planting seed that is infested with virus diseases or that is not in the right stage can depress yields substantially. The scarcity of reasonably priced, healthy seed tubers at planting time is the main factor inhibiting potato production in many developing areas.

Because potatoes are vegetatively reproduced, the technical and institutional requirements of a successful seed potato program are quite different from those for grains. Seed potatoes are more costly to multiply and distribute than are seeds of the cereals, pulses, and oilseeds. Often, seed of several crops can be handled by one processing and distribution facility. But potatoes and other vegetatively reproduced crops require specialized facilities and personnel. Because of the unique problems of vegetative reproduction, officials establishing national seed certification programs often decide to include grains but not potatoes and other root crops. Consequently, those who formulate national potato programs are compelled to establish mechanisms for multiplying healthy seed tubers and distributing new varieties.

Bibliographic notes

There is a vast technical literature on the potato crop, most of which is based on research conducted in northern latitudes under temperate growing conditions. Unfortunately many of the research results obtained under temperate conditions in developed countries have little direct application in the tropics and subtropics. Only a few of the most recent, significant, and accessible references are noted here. Additional sources on specific topics may be obtained from the Library, CIP, Aptdo. 5969, Lima, Peru.

Potato Abstracts, issued monthly by the Commonwealth Agricultural Bureaux, is the best guide to research on the potato. *Potato Research* (the quarterly journal of the European Association of Potato Research), *American Potato Journal* (published monthly by the Potato Association of America), and *La Pomme de Terre Française* are the leading international journals dedicated to publication of scientific papers on the potato. Potato journals are also published in several other developed and a few developing countries, including India and China.

Numerous potato production handbooks, textbooks, and collections of papers are available. Among the most authoritative are Beukema and van der Zaag (1979), Burton (1966), Harris (1978), Li (1985), and Smith (1977). Pushkarnath's work (1976) is the most comprehensive volume on potato production under subtropical conditions (particularly India). Montaldo (1964) provides an extensive, but now rather dated, bibliography on potatoes in Latin America. Ahmad (1977) persents useful information on potatoes in Bangladesh. Manuals on potato storage and processing have been published by Booth and Shaw (1981) and Shaw and Booth (1982), respectively. Tissue culture and genetic engineering of the potato are discussed in Espinoza et al. (1986) and Jaynes, Espinoza, and Dodds (in press).

The first section of this chapter on the potato plant draws heavily on Beukema and van der Zaag (1979) and Thornton and Sieczka (1980). The section on the physical environment is based on Beukema and van der Zaag (1979). The section on breeding new potato varieties is based on Mendoza and Sawyer (1985), Hawkes (1978a), Howard (1978), Huamán and Ross (1985), and on conversations with Carlos Ochoa, Zosimo Huaman, and Humberto Mendoza of the CIP staff.

3
The socioeconomic environment: Levels of development and public policies

Physical factors—climate, soils, pests, and diseases—influence the growth of the potato, but they do not determine where or how the crop will be grown or how the output will be used; social, economic, and political factors have a critical role as well. Humans intervene in the potato production process by manipulating both the plant and its physical environment. The complexity of these manipulations is illustrated by the sophistication of pre-Columbian Andean agriculture.

The Peruvian Andes presents one of the most diverse landscapes on earth. Its soils and climatic conditions vary dramatically over short distances. To fit the different local environments, early Andean cultivators selected a range of potato varieties. In the highest zones, at more than 4,000 meters above sea level, frost-resistant bitter potatoes (*S. juzepczukii* and *S. curtilobum*) were the only food crop that could survive. After being harvested and freeze-dried, the tubers could be easily stored, providing a welcome degree of food security in this harsh and risky environment. Higher yielding, late-maturing varieties with little frost resistance and long periods of dormancy were selected for intermediate altitudes (2,000 to 4,000 meters above sea level), where a single potato crop could be grown each year and seed potatoes had to be stored for 5 or 6 months. Early-maturing varieties were selected for lower, warmer areas where potatoes could be grown at any time of the year. These varieties generally yielded less than the later varieties grown in higher areas, but their earliness and short dormancy allowed farmers to grow potatoes in sequence with maize and other crops and to replant potatoes soon after harvest.

Early Andean farmers also modified the physical environment in important ways. Irrigation canals and stone-walled terraces were constructed throughout the Incan empire, converting barren mountainsides into productive cropland for maize, potatoes, and other crops. Agricultural implements, such as the Andean foot-plow (*chaquitaclla*), were devised for land preparation, weeding, and harvesting. Livestock manure was applied to improve soil fertility and structure. Pests, such as the cyst nematode, were controlled through use of communal rotations and long fallow periods, with potatoes being grown at most only once every 7 years.

In Peru, as elsewhere, traditional agriculture has been altered by population growth, urbanization, expansion of markets and increasing specialization in production, the advance of scientific knowledge, and the growing influence of government policies and programs. In modern times, the Andes of Peru have become linked to national and world markets, giving rise to commercial potato production based on purchased inputs. Although farmers in remote areas continue to grow native varieties for their own consumption, most sell part of their harvest and purchase some inputs. The process of varietal selection has been accelerated by Peru's Ministry of Agriculture, which has been releasing new varieties for 30 years. New crops, like rice and wheat, have been introduced, and food imports have expanded dramatically. Trade policies intended to help Peruvian industries have had the effect of overvaluing Peru's currency on international financial markets and reducing the price of imported food in Peru. As a consequence, the competitive position of the potato crop has been gradually eroded. Urbanization and increases in per capita income have also stimulated households to diversify their diets and consume smaller quantities of traditional staple foods like potatoes. The end result has been shrinking per capita potato consumption, a decline in subsistence-oriented production, and a gradual shift of potato production to favorably endowed areas with good access to urban markets.

The development of potato production in India provides a striking counterpoint to the pattern in the Andes. For several centuries after their introduction, potatoes were grown only in the northern hill areas and in small pockets near large cities. The difficulties of producing and marketing potatoes made them a costly delicacy in most of the Indian subcontinent. Consumption was limited to wealthy Indians and foreigners. Even with time, as incomes rose and more people could afford to eat potatoes, consumption levels remained low until production and marketing costs were brought down.

Population growth and urbanization have expanded the market for food crops. The spread of tube-well irrigation and introduction of earlier varieties of cereals have allowed potato production to expand in the populous Indo-Gangetic plain in the winter season. An effective seed production system was developed for the plains, and the release of new varieties raised yields, reducing unit costs of production. The construction of cold storage facilities allowed farmers to hold both seed and consumer potatoes through the hot summer months. These technological improvements and economic changes have combined to make the potato industry one of the most dynamic subsectors of Indian agriculture.

Economic development

The transformation of a subsistence-oriented, agrarian economy into a monetized, urban-based, capital-intensive economy with expanding service and industrial sectors has many effects on agricultural production. It is no coincidence that Europe's industrial and agricultural revolutions took place simultaneously.

Four aspects of contemporary economic development can be expected to affect potato production and use: increased per capita income; urbanization; improvements in transportation; and declining relative prices for agricultural inputs such as tractors, implements, fertilizers, and pesticides. Rising incomes stimulate demand for exotic and therefore costly foods to diversify diets. This trend is most pronounced in the cities, where consumers tend to emulate foreign dietary patterns. Hence, if potatoes are relatively expensive, as they are in most developing countries, rising incomes can be expected to stimulate the demand for them. On the other hand, where potatoes are a cheap, traditional staple food, as in the southern Andes and temperate zones, income growth may decrease the demand for potatoes.

Better transportation between rural and urban areas stimulates potato production and consumption by helping to reduce the marketing costs for bulky perishables. Declining prices of purchased inputs relative to prices for farm products give farmers an incentive to expand production of crops like potatoes, the yields of which are highly responsive to improvements in tillage and irrigation as well as to greater application of chemical fertilizers and pesticides.

General economic development is intrinsically linked to the expansion and integration of markets. In subsistence economies, most agricultural inputs—land, labor, manure, implements, and draft animals—originate on the farm, and most of the output is used by the farm household. For these reasons, farmers in areas that are isolated from input and product markets grow crops that are well adapted to the natural environment, that produce well without purchased inputs, and that are desired for home consumption and home manufactures. Farms tend to be small and highly diversified. Crop intensity and yields are usually low, and a substantial part of the farmland is in pasture and fallow. A household's supply of labor and draft animals—the principal sources of energy for cultivation—usually limits the size and intensity of its farming operation. Farmers attempt to maximize their productivity by diversifying activities and spreading the use of energy over time. Diversification also lessens the chances of total crop failure due to weather or pest problems, which is the farmer's major risk. Potatoes are generally grown in subsistence economies because for growers they provide a cheap source of calories for human consumption or livestock feed.

As economies develop, farmers tend to concentrate on fewer crops or those types of livestock for which they have a comparative advantage. They market more of their output and use the income to buy food, clothing, and other consumer and capital goods. Tractors begin to replace labor and draft animals as the main source of power. Fallows are shortened, cropping intensity is increased, land improvements are made, and chemical fertilizers and pesticides are applied to increase and stabilize yields. The levels of productivity and value added per hectare rise. However, cash returns and net incomes often fluctuate substantially from year to year in response to changes in prices and output. In market economies, potatoes may be produced because they are a cheap source of energy or because they can be profitably marketed as a delicacy.

It is impossible to predict the precise effects of economic development on potato production. However, a few important generalizations can be made. As markets expand, it becomes profitable to purchase yield-increasing inputs like fertilizers and pesticides. Market expansion also opens the way for specialization in production. The first outcome may be that a few large commercial potato farms dominate the market, with small farms growing potatoes only for home consumption; or it may be that many small farmers grow potatoes for market using hired draft power and purchased agricultural chemicals. The specific route taken depends in large part on government policies. There are strong advocates of both positions. Where potato production has been restricted by unfavorable growing conditions, inappropriate technology, costly inputs, and limited markets, economic development can be expected to reduce costs and stimulate potato production and consumption. This situation can be considered typical in most developing countries. On the other hand, where potatoes are already a low-cost staple food because they grow well in the local environment with minimal use of purchased inputs, economic development is likely to erode their competitive position relative to higher value crops that respond well to purchased inputs and can now be sold profitably in distant markets. One reason is that as incomes rise, consumers diversify their diet away from the basic staples. Moreover, improved market integration reduces the price of foods imported from other areas. This happened in Ireland a century ago and is taking place in some highland and temperate zones of developing countries today. The final generalization is that rural population growth, which results in declining farm size and increased land prices, stimulates the cultivation of high-yielding crops such as potatoes.

Potatoes are an important cash crop in northern Ecuador.

Government policies and programs

Government policies and programs influence potato production and use in many ways, some of which are unanticipated or unintended. The effect of general policies and programs that are designed to achieve broad social, economic, or political goals are reviewed in this section. Programs and policies that are aimed at goals specifically related to potatoes are discussed in later chapters.

Macroeconomic policies and development strategies

Government policies and programs that are far removed from agriculture can stimulate potato production or hinder it. Government programs can also influence the course of technological change and the distribution of its benefits among different social groups.

Policies that change the ratio between potato prices and those of inputs and other agricultural commodities affect the profitability and growth of the potato subsector in relation to agriculture as a whole. For example, politically powerful urban groups often press the government to hold down food prices, and industrialists may lobby for cheap-food policies to help them keep wage rates low. Where urban groups succeed in enforcing ceiling prices for potatoes, as has occurred from time to time in Andean countries, farmers lose interest in expanding potato production, causing prices to go even higher eventually.

Trade restrictions and foreign exchange policies may also discourage potato production. To protect local industry, many governments impose tariffs and other barriers to importation. These trade barriers limit the local demand for foreign currency and cause the country's currency to be overvalued in international markets. This, in turn, lowers the price (in domestic currency) of imported foods—primarily wheat and rice. Consequently, protection of manufacturing and overvaluation of exchange rates has the effect of taxing agriculture and reducing the competitive position of domestically produced foods, such as potatoes. Slow agricultural growth increases the country's dependence on imported food. Given the tendency of many developing countries to overvalue their exchange rates, and in effect subsidize food imports, the rapid increase in potato consumption in these countries is quite surprising.

Policies that affect the availability and price of agricultural inputs may also influence production methods and yields. In Rwanda, for example, chemical fertilizers and fungicides are scarce because they are not manufactured locally, and foreign exchange is not made available for importing them. Yet, on-farm trials in Rwanda show that under a wide range of prices, use of fungicides on potatoes would be highly profitable. A policy that facilitated the importation of fungicides would stimulate potato production and decrease crop failure in the rainy season.

Similary, in many developing countries, a policy that increased lending to farmers and lowered the real rate of interest (the current rate minus the rate of inflation) would encourage use of purchased inputs on potatoes. Also, allowing farmers more time to repay their loans would stimulate potato storage, helping to moderate seasonal fluctuations in market supplies and potato prices. Usually, farmers' loans come due at harvest time, forcing them to sell their harvest when prices are lowest.

Marketing policies can affect potato production and use in many ways, some of which are discussed in the next chapter.

Land tenure

Potato production and marketing systems, as well as research and extension, need to be geared to the existing land tenure system, and changes in tenure patterns require corresponding adjustments in these systems. Large-scale potato farms can afford to perform a number of marketing, credit, research, and extension functions for which small farms must rely on the public sector or farmers' associations.

In the Cañete Valley on Peru's coast, for example, prior to the land reform of 1968, the transportation, marketing, irrigation, research, and extension systems had all evolved to meet the needs of a few large farmers. The land reform initially transformed large private farms into production cooperatives; later, they were subdivided into many small private holdings. As a result, the valley's irrigation system needed drastic restructuring. Large-scale equipment had to be replaced with smaller models. The government also found it necessary to greatly expand the extension service to meet the needs of the valley's large number of new small farmers. Credit and marketing arrangements had to be changed, too. The transformation from large private farms to production cooperatives to small private farms was accompanied by a decline in potato production and yields. In recent years, these trends have been exacerbated by unusually warm weather during the growing season and a dramatic increase in insect populations, which have cut yields even further.

Because different types of farmers—small subsistence growers, large commercial farmers, market-gardeners, production cooperatives—have different technological requirements and place different demands on services such as credit, marketing, and irrigation, it is important that researchers and policymakers act with awareness of local land tenure situations.

Regional development priorities

Government policies, investments, and development programs usually favor some regions over others for various social, economic, and political reasons. These regional priorities can have important effects on potato production and use. In Mexico, for example, potato production has expanded most rapidly in the Northwest, where the country's major irrigation projects are located. In Bangladesh, potato production near Dhaka has been stimulated by construction of cold storage facilities. In Peru, potato production has expanded most in the coastal valleys, which have benefited disproportionately from public investment.

Given the high costs of transporting potatoes, building a road into one area can dramatically improve that area's competitive position relative to other areas. Potato production has followed road construction into mountainous areas in several parts of Africa and Asia in recent years. If a potato-growing area is low on the government's list of regional priorities and does not benefit from road construction, agricultural credit, rural development projects, and other institutional support, it is unlikely that potato research and extension efforts will be able to accomplish much in terms of increasing production or rural welfare.

Agricultural research policy

Research policy is probably the single strongest instrument the public sector can employ to directly influence long-term trends in crop production. New technologies that lower production and marketing costs can enhance the competitive position of the potato crop and provide incentives for expanding production, consumption, processing, and trade. Research priorities and strategies are discussed in Chapter 7, and their impact on potato production and use in Chapter 8.

Bibliographic notes

The most authoritative reference on the topics covered in this chapter is *Agricultural development,* by Hayami and Ruttan (1985). Other important references include Mellor (1966), Johnston and Kilby (1975), and Eicher and Staatz (1984). The footnotes of Hayami and Ruttan's text contain additional useful references.

4
Supply, demand, and marketing

Most potato growers derive cash income from the crop. Where potatoes are a high-priced food, farmers usually market their harvest and use the income to purchase less expensive staples for their own needs. Even where potatoes are cheap enough for farm families to consume as a staple food, farmers usually sell part of their crop.

Potato prices in most countries are set by market forces rather than by government fiat. Changes in price levels strongly influence both the decisions and the welfare of potato producers and consumers. Thus marketing is not only important but controversial. Producers usually feel the prices they receive for their potatoes are too low, whereas consumers feel the prices they pay are too high. Moreover, it is often believed that low producer prices, high consumer prices, and price instability are due to inefficient or exploitive marketing practices. This is sometimes, but not always, the case.

Supply and demand

The branch of economics known as price analysis provides a useful framework for understanding how prices are determined and for pinpointing marketing problems. The concepts of supply and demand are at the heart of price analysis. A *supply curve* indicates the quantities that would be offered for sale or supplied by producers at different price levels; a *demand curve* indicates the quantities that would be demanded by consumers at different prices (Fig. 7). When supply and demand curves are plotted on

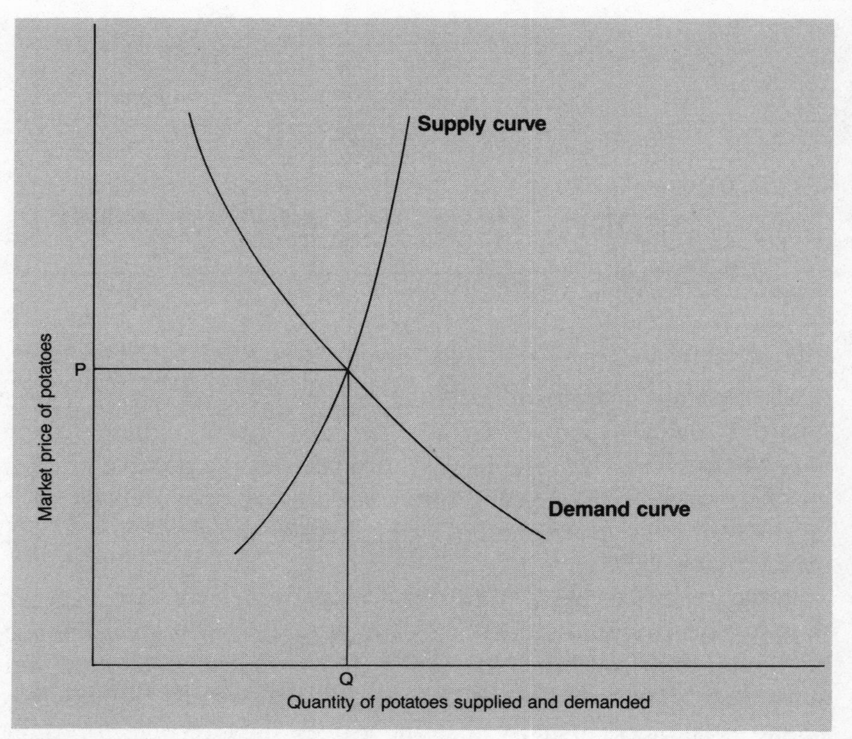

Figure 7. Supply and demand curves, showing the market equilibrium price (P) and quantity (Q).

the same scale, their point of intersection marks the equilibrium price at which the amount offered for sale is equal to the amount that will be purchased.

The principles of supply and demand apply in both market economies and centrally planned economies. When governments fix prices, signs that the government-mandated price is below the market equilibrium level are shortages, rationing, and long lines of would-be customers standing outside food stores. Stockpiled surpluses, on the other hand, indicate that prices are above the equilibrium level.

Supply

Over time, the supply curve for potatoes may shift in response to a variety of changes: in the prices of inputs used in potato

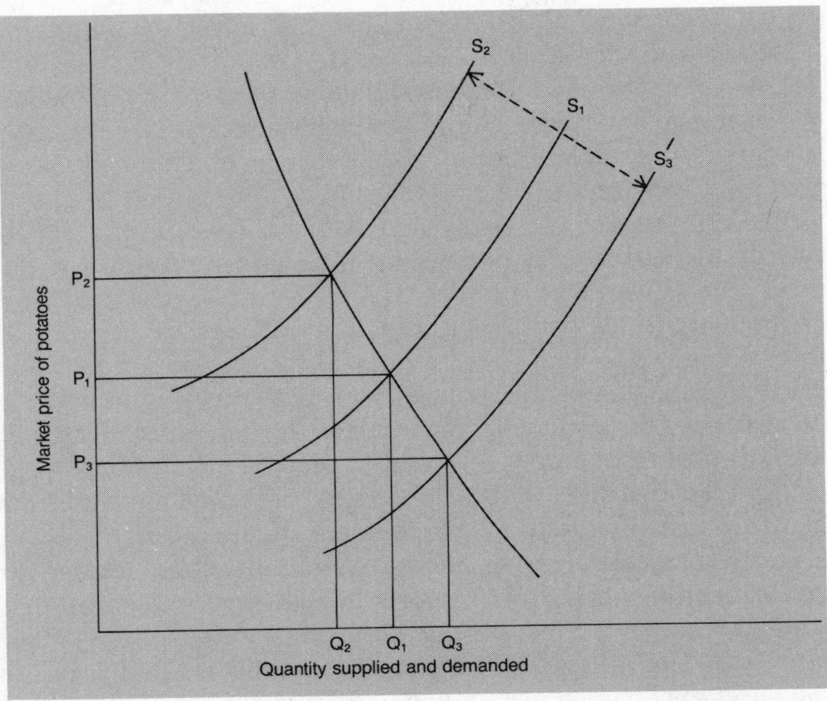

Figure 8. Effects of a shift in the supply curve on the price and quantity of potatoes supplied and demanded.

production; in the production technology employed by farmers; in the supply of land, labor, and capital; and in environmental conditions. If input prices increase, for example, the supply curve will shift upward and to the left (from S1 to S2 in Fig. 8). Producers will supply fewer potatoes to the market (Q2 rather than Q1), and prices will rise (from P1 to P2). On the other hand, if input prices fall, the supply curve will shift downward and to the right (from S1 to S3). Farmers will supply more potatoes (Q3 rather than Q1), and prices will fall (from P1 to P3). If farmers begin to use new technologies that enhance the productivity of inputs, the supply curve will shift to the right, enlarging market supplies and reducing prices.

In the short run, the supplies of many factors of production, such as land, equipment, and labor are fixed; but over time they can also change. Changes in the amount of land, labor, and capital devoted to potato production will affect the supply of potatoes, and vice versa.

Environmental conditions also influence the supply of potatoes. For example, bad weather or a late-blight attack may reduce the supply of potatoes and boost prices, whereas exceptionally good growing conditions will increase supplies and depress prices.

These principles of supply and demand have several implications for agricultural policy, research, and extension. In the short run, government policies that make inputs like seed and credit more readily available and that reduce their costs will induce farmers to supply more potatoes to the market at lower prices. On the other hand, policies such as restrictions on imports of fertilizers or pesticides that result in higher input costs will tend to limit the supply of potatoes and raise the price. Agricultural research and extension programs that bring improvements like new varieties and pest management systems to farmers will also encourage farmers to produce more potatoes, leading to greater supplies and lower prices. Policies and programs that stimulate capital formation and land improvements on potato farms will lead to gradual expansion in supplies. Finally, policy measures may help moderate fluctuations in market supplies and potato prices caused by weather and other environmental conditions, but they are not likely to eliminate them altogether.

Supply response refers to movements *along* the supply curve, in contrast to *shifts* of the supply curve discussed earlier. A conventional measure of supply response is the *price elasticity of supply,* which is defined as the percentage change in the quantity supplied that results from a 1 percent change in the market price of the commodity. The precise magnitude of the price elasticity of supply depends on a number of factors that vary from place to place. Highly reliable estimates are not available for potatoes in most developing areas, but several useful generalizations can be made.

First, the degree of supply response increases with time because factors of production that are fixed in the short run, like land and labor, become variable in the long run. Thus, if greater demand for potatoes forces prices higher, farmers are likely to devote more land to potatoes and invest more money in inputs.

Second, the supply response for any single crop is greater than the supply response for all crops together because farmers can easily react to changes in prices by shifting their resources from one crop to another, but their aggregate supply response is limited by the total supply of land, labor, and capital.

Third, farmers' supply responses are greater in economically advanced areas, which have well-developed markets for inputs and products, than in more backward areas. In the more prosperous areas, farmers are more market-oriented, and purchased inputs play a greater role in their farming systems. As a result, a country may develop a "dualistic" structure of potato production: one highly market-oriented and dynamic, the other subsistence-oriented and relatively static.

Fourth, among the food crops, supply response is greatest for crops like potatoes, whose yields are very responsive to variable inputs like labor, fertilizers, pesticides, and management.

For these reasons, in most countries the area and production of potatoes fluctuate widely from year to year as farmers adjust their cropping and input levels in response to changing market conditions. An additional source of short-term supply variations is the fluctuation in yields caused by changes in weather, pests, and diseases.

Demand

The demand for potatoes reflects biological needs, food habits, income levels, and prices.

Biological needs and food habits. Although food consumption is essential for human growth and maintenance, eating is much more than a biological process. Food consumption patterns are strongly influenced by culture and habits. Early in life, people develop food preferences, which usually stay with them throughout life. Asian students in Europe, for example, continue to prefer rice over potatoes, whereas Europeans residing in the tropics prefer potatoes over yams or rice.

The conservative nature of food habits has had a profound effect on potato consumption trends in different parts of the world. In 1850, when massive European migration to the United States began, European levels of potato consumption were more than double those of the United States. The immigrants continued to eat many potatoes, even though maize and wheat were cheaper calorie sources in North America. The result was a significant growth in per capita potato consumption in the United States in the last half of the nineteenth century. Later, as the immigrants gradually brought their eating habits into line with American customs and food prices, potato consumption levels fell. The advent of "fast foods" often eaten with french fries has spurred potato consumption once again.

The resistance to change that is characteristic of food habits has sometimes limited potato consumption in developing countries; nevertheless, significant dietary changes do occur over time. In countries as culturally distinct as Germany and Rwanda, potato consumption was initially restricted because people thought potatoes were unhealthy. When other foods became scarce due to crop failure and political disturbances, people learned that potatoes were a palatable and nutritious food. In essence, the demand curve for potatoes shifted upward and to the right. After these events it never returned to its original lower level.

Food habits can also change through the demonstration effect. As regions become integrated into a global economy and society, food habits become more universal. Irrespective of price, Europeans begin to eat more rice, for example, and Asians begin to eat more potatoes. Urbanization and changes within the household economy also influence food habits. As women begin to work outside the home and have less time for food preparation, more meals are eaten in restaurants and institutions (such as schools), and consumption patterns shift toward use of precooked and fast foods.

Income. Households vary the composition of their diets in response to changes in their income. Low-income consumers rely mostly on cereals, root crops, plantains, or pulses for calories and protein. As their incomes rise, they diversify their meals by consuming larger amounts of higher-priced fruits, vegetables, meats, oils, and dairy products.

Engel's law, one of the most general of all economic relationships, relates income to the proportion of total expenditure used for food: *the smaller the family income, the greater is the proportion of family income that is spent on food.* A second

TABLE 8
Income Elasticities of Demand for Potatoes in Relation to Per Capita Income

	Demand elasticity	GNP per capita 1980
Developed market economies	−0.21	9,610
North America	−0.19	11,243
Oceania	−0.10	9,314
Europe	−0.19	9,059
USSR and Eastern Europe	−0.33	4,475
Developing market economies	0.19	773
Latin America	0.12	1,831
Near East	0.11	1,617
Africa	0.51	611
Far East	0.16	371
China and other Asian centrally planned economies	0.30	268

Source: FAO, Commodity Policy and Projections Service, Commodity and Trade Division (1984, unpublished); and D. Horton and H. Fano, *Potato atlas* (Lima: International Potato Center, 1985).

relationship, formulated by Merrill K. Bennett in the 1930s, relates income level to the composition of the food basket: *the smaller the family income, the greater is the relative contribution of starchy staples to total calorie intake.* Bennett also observed that within the broad category of the starchy staple foods, consumers tend to switch from roots and coarse grains to the preferred cereals, usually wheat or rice, as incomes rise. On the basis of these observations, which have proven to have broad validity in different times and locations, it has sometimes been concluded that as incomes increase, per capita consumption of potatoes will fall. This is not necessarily true. It all depends on the level of income, the relative price of potatoes, and whether potatoes are chosen primarily as a calorie provider or for their flavor and texture.

The response of consumption to income changes is measured by the *income elasticity of demand,* which is defined as the percentage change in the quantity demanded that results from a 1 percent change in per capita income. In poor areas, income elasticities of demand for foods, including potatoes, tend to be higher than in rich areas. In fact, income elasticities of demand for fresh potatoes (as opposed to processed potatoes) tend to be negative in developed countries and positive in developing countries (Table 8). That is, as incomes rise in developed countries, fresh potato consumption falls; whereas in developing countries, rising incomes lead to rising consumption. Also, within any country, income elasticities of demand for potatoes tend to be higher for the poor than for the rich (Table 9).

These relationships have important implications for future potato consumption. First, in most developing countries, gains in national income will expand demand for potatoes. (This is not always true in the developed countries.) Second, within a country, an increase in the income of the poor will ordinarily have more impact on the demand for potatoes than an increase in the income of the rich. Finally, the manner in which demand for potatoes responds to changes in income depends on the price of potatoes and the role of potatoes in the diet. For example, where potatoes are so expensive that poor people do not eat them at all, an increase in the income of the poor may have no effect on the demand for potatoes.

Price. The long-term effects of changing food prices on consumption patterns have rarely been studied. It is clear, however, that prices have a strong influence on consumption decisions in developing countries. Consumers in poor areas choose locally abundant, cheap foods as their major source of energy. These are mainly starchy roots or cereals and pulses that are well adapted to local growing conditions. As local availability and food prices change over time, so do consumption patterns. In Europe, when potatoes became the cheapest source of energy, for example, they replaced oats, buckwheat, and other cereals in the diet. Similarly, in parts of Africa, cassava has replaced yams and cocoyams; and in China, sweet potatoes have become a major staple food. All these crops, introduced from the Americas, replaced other more costly indigenous food crops.

TABLE 9
Estimated Income Elasticities of Demand for Potatoes by Income Group in Selected Countries

Countries	Income elasticity of demand by income group			
	Lowest income	Second lowest	Third lowest	Highest income
Developing countries				
Bangladesh (nationwide, 1973-1974)	0.32	7.11	0.23	0.31
Chile (Gran Santiago, 1977-1978)	0.63	−0.14	−0.17	−0.14
Panama (nationwide, 1981)	0.68	0.27	0.45	−0.03
Senegal (Dakar, 1975)	6.24	0.91	0.47	0.35
Somalia (Mogadishu, 1977)	4.59	9.05	6.97	2.06
Sri Lanka (Colombo, 1977)	0.48	1.09	1.42	1.04
Zimbabwe (urban areas, 1975-1976)	−0.11	0.97	0.56	−0.16
Developed countries				
Australia (Sydney, 1974-1975)	−0.16	0.00	0.00	−0.14
Belgium (nationwide, 1974-1975)	−0.43	0.36	0.37	−0.27
Canada (urban areas, 1976)	0.75	−0.70	−0.49	−0.28
Greece (nationwide, 1974)	0.43	0.50	0.22	−0.02
Ireland (towns, pop. 1000 or more, 1979)	0.00	−0.09	−0.03	−0.21
Poland (nationwide, farmers, 1976)	0.24	0.08	0.08	0.00
United Kingdom (nationwide, 1980)	−0.10	−0.18	−0.50	0.18

Source: Derived from FAO, *Review of food consumption surveys 1981* (Rome, 1983).

Insofar as they can, people prefer to eat a variety of foods rather than a single staple. Within a fixed budget, they purchase different quantities of various foods, depending on their prices. They buy more of a food when it is cheap than when it is expensive, but even when it is costly they rarely choose to forgo its consumption altogether. If potato prices rise, people consume fewer potatoes and increase their intake of cheaper foods. But even when potatoes are expensive, many consumers still purchase a few to eat, perhaps on special occasions.

The relationship between price and the quantity demanded is reflected in the demand curve for a commodity. The degree to which people increase or decrease their consumption of a food in response to a change in its price is measured by the *price elasticity of demand,* which is the percentage change in the quantity of a food demanded that results from a 1 percent change in its price. Because a price reduction typically results in greater consumption (and the converse), the price elasticity of demand is usually a negative number.

It is important to know how consumers respond to changes in the price of potatoes. Although, unfortunately, few economists have studied the demand for potatoes in developing areas, it is clear that in a locality, the absolute value of the price elasticity of demand for potatoes rises the longer the price is in effect. The practical implication is that a sudden expansion in the supply of potatoes is likely to cause prices to fall sharply, but if the higher level of supply is sustained, prices will creep up again as consumption grows. In the very short-term (a few days or weeks), people do not normally switch between different staple foods in response to price changes. Unexpected increases in supply oftentimes lead to sharp price drops due to the high cost of storing and transporting potatoes. If the potato price remains low for several seasons, however, consumers can be expected to respond by eating more potatoes more often. The greater the number of alternative uses and substitutes for a food, the more sensitive demand is to price. For this reason, the price elasticity of demand for potatoes depends on food habits and the prices of competing foods, feeds, and industrial inputs. Because the poor have greater difficulty meeting their food requirements and spend a higher proportion of their income on food, they are more responsive to food prices than are more affluent groups. Thus, price elasticities of demand for potatoes are generally higher in poor countries than in rich ones.

In many countries, marketing problems have accompanied rapid increases in potato production. Occasional market gluts and spoilage of large volumes of potatoes have marred the otherwise favorable image of potato programs in countries like India, Bangladesh, and Pakistan, where, due to lack of adequate storage and transportation facilities, bumper harvests have resulted in temporary local surpluses. However, these short-term marketing problems have not discouraged further increases in production. Wherever introduction of improved potato technology has increased supplies, demand has also grown. Despite occasional marketing problems, potato production has more than tripled in several developing countries over the last two decades, and potato production continues to be highly profitable for most farmers. Cost reductions in potato production and marketing have allowed farmers to expand production and increase their profits at the same time that prices to the consumer have fallen.

Marketing and prices

In a purely subsistence economy, every household consumes what it produces and there is no marketing. Yet what are conventionally considered marketing functions are still performed: The farm family transports foods and other agricultural products from the field to the housing compound, stores them from harvest time until they are needed later, and transforms them into products that keep better or are considered more desirable for household consumption. These activities must be financed in the sense that scarce resources (labor, building space) must be allocated to them. And decisions must be made about the distribution of the harvest within the household.

But most potato growers, even in isolated places, sell at least part of their output. For them, marketing may be more critical than production because a sharp drop in price can cancel out all the potential benefits of a bumper crop. At harvest time, if all potatoes in a region are offered for sale, the market supply mounts abruptly and the price plummets. After the harvest, the market supply of potatoes contracts and the price rises again. In countries like Bangladesh, prior to the introduction of cold storage, potato prices crashed each year at harvest time; a few months later, prices soared as potatoes virtually disappeared from the market. Storage has helped moderate price cycles in Bangladesh but has not eliminated them altogether (Fig. 9). At harvest time, February and March, potatoes cost a third as much as in October and November (Fig. 10).

Marketing strategies

Producers can take various steps to protect themselves from the collapse of prices at harvest time. Among them are off-season production, transportation to other markets (domestic or foreign), storage, processing, and feeding surplus potatoes to livestock. Aside from off-season production, these activities are generally thought of as *marketing functions* and may also be performed by market intermediaries or sometimes by consumers.

Off-season planting. Some farmers plant a small area of potatoes before or after the main season in order to take advantage of high off-season prices. The drawback is that early or late crops usually are more costly to produce than the main crop, and their yields are commonly lower and more variable. Nevertheless, farmers often find that staggered planting and harvesting are more economical than storage, processing, hauling potatoes to other markets, or feeding them to livestock. One action policy-makers can take to encourage off-season production is to ensure that credit is available when farmers need it. Researchers and extension workers can help by giving priority to the major technological constraints such as the shortage of seed tubers for off-season planting or the lack of suitable fast-growing varieties. Finding ways to overcome such problems can go a long way toward smoothing fluctuations in potato supplies and prices.

Domestic marketing. The bulkiness and perishability of potato tubers makes them costly to ship from areas of surplus production to distant markets where prices are higher, especially if the production takes place in mountainous areas and the urban centers are in the lowlands. Yet if price differentials between

Near Agra, India, workers sort potatoes that will be placed in cold storage.

markets are large, if transportation costs are low, and if losses during shipping can be minimized, potatoes will be moved long distances.

To induce producers or traders to ship potatoes, the price differential between markets must be greater than the cost of shipping potatoes. For this reason, intermarket price differences are largest where transportation costs and losses are highest. Sometimes local governments ban shipments of potatoes to other provinces or states in order to lower the local price to consumers. Policymakers can help improve domestic potato marketing by eliminating such barriers to the movement of potatoes. Finding

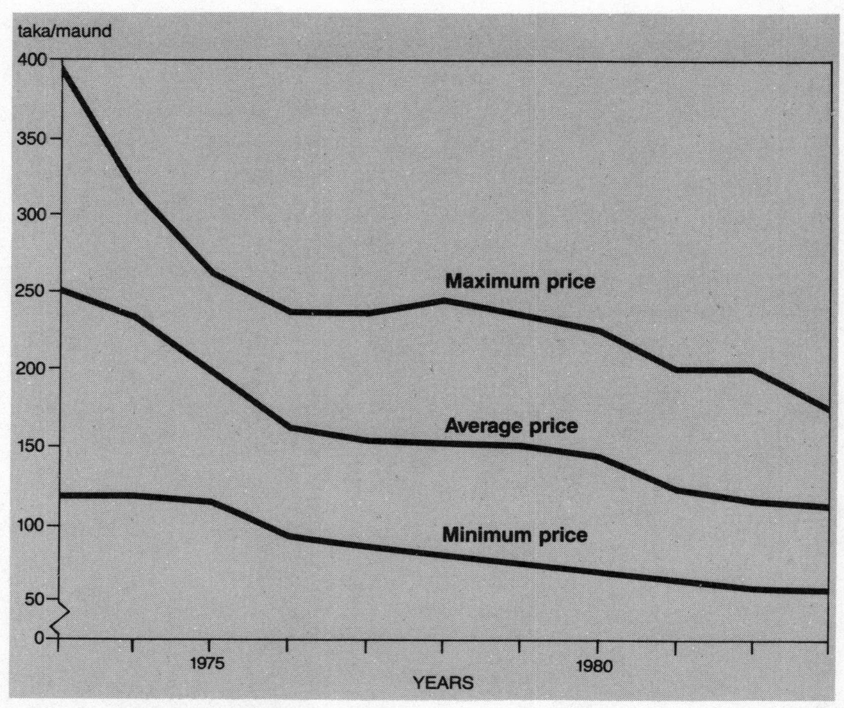

Figure 9. Bangladesh: Three-year moving average of wholesale potato prices (deflated by index of wholesale food prices: 1984 = 100). ***Source:*** **Data provided by Directorate of Agricultural Marketing, Dhaka.**

means of reducing transportation costs and losses, particularly in warm areas, should be a high priority for researchers and extension workers.

Foreign trade. Exporting often seems an alluring way to expand the market for potatoes. However, in most countries only a small fraction of the potato harvest is exported because of the potato's bulkiness and perishability. Exporting requires specialized equipment and efficient coordination to avoid high transit losses. But unless supplies available for export are reliable, the refrigeration needed in warm areas for hauling and storing potatoes is not economical, Also, some potential markets impose barriers to imports. Quarantine systems may restrict imports of vegetative materials like potatoes, which can carry diseases. Local producer

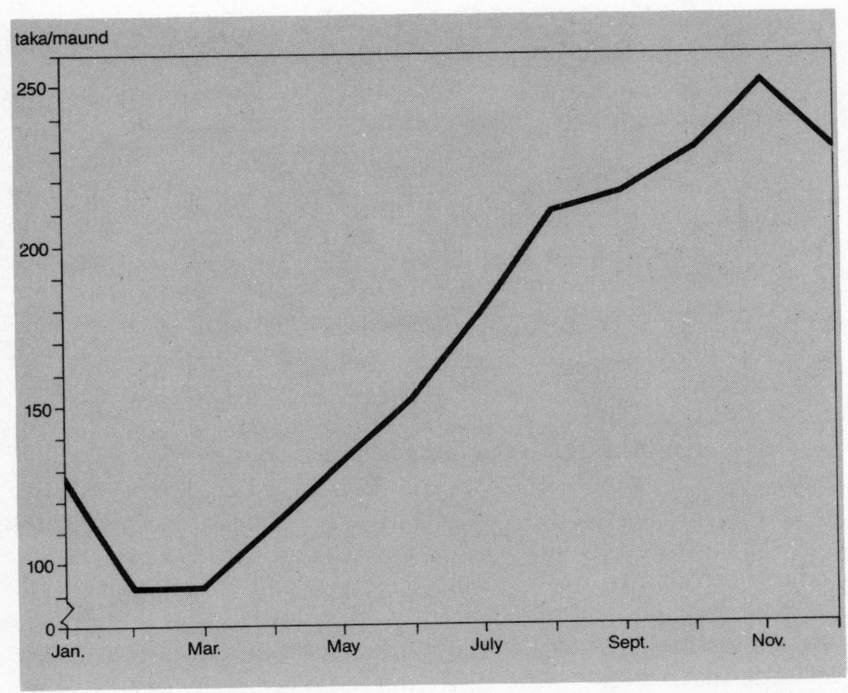

Figure 10. Bangladesh: Seasonal cycle of retail potato prices (average monthly prices 1972–84 deflated by the consumer price index—1984 = 100). *Source:* Data provided by Directorate of Agricultural Marketing, Dhaka.

groups often press government agencies to ban imports on the grounds that foreign potatoes bear diseases. This argument can be valid, but is sometimes simply an excuse to keep cheap potatoes out of the market.

An important distinction must be made between cereals and potatoes in international trade. While world markets for cereals are highly organized, potatoes are customarily traded regionally and on a more personal basis. The high cost of shipping potatoes (relative to the price of the commodity), the instability of supplies and prices, and the heterogeneity of potatoes traded (different varieties, size grades, physiological condition, and health) prevent the operation of a broad world market with generally accepted standards and impersonal transactions. Consequently, agencies attempting to develop potato exports must do more careful market research and establish more personal trading arrangements than would be required to initiate exports of cereals.

Despite the problems, a well-managed potato trade can be lucrative. Taiwanese traders, for example, profit handsomely from supplying potatoes to Hong Kong and Singapore. Guatemala, because of its highland growing areas and low production costs for potatoes, exports to neighboring countries. Similarly, Rwanda exports to neighboring Burundi and Zaire. Developing countries in the Mediterranean region, such as Cyprus, Egypt, Lebanon, and Morocco, export significant volumes of potatoes to Europe. Many of them have established reciprocal trading arrangements: They export consumer potatoes to countries from which they import seed potatoes.

Owing to substantial reductions in production costs and potato prices in countries like India and Bangladesh recently, exports to nearby markets could perhaps generate foreign exchange and stimulate greater potato production. In fact, in the late 1970s, Indian merchants were developing profitable Middle Eastern markets; but political pressure from consumer groups in Bombay caused the national government to ban potato exports on the grounds that exports were depleting local supplies and inflating the price of the "poor man's vegetable."

A sensible first step towards expanded trade would be to remove legal barriers to exporting. Then, seasonal supplies and price movements would need to be studied, and major marketing costs and risks identified. Before investing in refrigerated railway cars, warehouses, and shipping containers, small experimental shipments should be made to confirm the profitability of trading potatoes. Usually, the best role for government is to facilitate private potato trade rather than become directly involved in commercial activities.

Storage. In many areas, farmers can profit by withholding potatoes from the market at harvest time and selling them later when prices are higher. But the bulkiness and perishability of potato tubers makes storage expensive and risky. The interest on the money tied up in stored potatoes is an additional cost. (Interest should also be considered when potatoes are moved to other markets, but because this is usually done rapidly, interest costs are much lower.) In areas where the harvest is followed by cool weather, potatoes can be held in simple structures for several months with minimal losses. But in lowland tropical areas where potatoes must be stored during the hot season, losses may be high unless they are placed in costly refrigerated structures.

Producers or market agents can afford to store potatoes only if the expected price rise will cover storage costs and risks. If the potato harvest is concentrated in a short period each year and prices advance predictably after harvest, storage may be profitable even when the costs or losses are high. In temperate zones, consumer potatoes are commonly stored because the costs and losses are relatively low and the harvest is concentrated in a few months, after which prices rise predictably. In India, Bangladesh, and Pakistan, it is profitable to store potatoes under refrigeration because the high costs are more than covered by the large, predictable increase in potato prices that occurs after each spring harvest (as shown previously in Fig. 10). Storage is less common in highland areas where potatoes are harvested throughout the year and no clear seasonal price cycle prevails.

Government agencies often erroneously assume that potato price instability can be overcome by erecting large storage facilties. Usually, however, the high costs of storing potatoes and the spoilage losses that occur are not covered by the price increase. As a consequence, government-built facilities rarely are used to store potatoes for more than a few seasons.

Instead of taking direct responsibility for storage, policymakers can improve potato marketing by making it attractive for farmers and market agents to store more potatoes. In some areas, storage is considered an antisocial or speculative activity and is discouraged by fines or forced sales in times of scarcity. These measures, by raising the costs and risks of marketing, expand the price spread between farmers and consumers and discourage potato production and consumption. On the technological side, researchers can contribute to market efficiency by finding ways to lower storage costs and losses on farms. In hot climates, evaporative cooling systems may be more cost-effective than refrigeration, especially for short-term storage. It may also be possible to identify new varieties that keep better than those currently used.

Processing. Through processing, potatoes may be converted into products that are less bulky and perishable and less expensive to store and transport. In the United States, over half the potatoes consumed are processed; most are pre-cut, frozen french fries. Research on processing and construction of low-cost facilities during World War II set the stage for greatly expanded potato processing in the United States. A mass market for processed potatoes was created when more women began to take jobs outside the home. Rigid grading standards and the availability of cheap, low-grade potatoes also stimulated U.S. processing.

In developing areas that do not possess that combination of factors, increased processing cannot be expected to greatly expand the demand for potatoes or solve marketing problems. Consumption of french fries is expanding rapidly (from a very small base) in many urban areas, as middle and high income consumers emulate foreign consumption habits. However, these and most other forms of processed potatoes are likely to continue to be high-cost foods consumed in relatively small quantities. An additional obstacle to potato processing is that the equipment is usually costly, imported, and complex. Any substantial investment in potato processing facilities should be preceded by market studies. The aim should be to find ways to produce desired products with facilities that can be built with local, rather than imported, materials.

Livestock feeding and industrial use. Potatoes are grown for feeding to livestock or use as the raw material for starch or alcohol production only where potatoes are cheaper than alternative sources. That situation prevailed in Europe a century ago, but it is uncommon in most areas today. Small, damaged, or spoiled potatoes, however, are often used for simple processing or livestock feed. Of course, in years of bumper crops when prices are low, the proportion of the harvest that is processed or fed to livestock is greater than in years of low production and high prices.

Raw potatoes may be fed to cattle, but for pigs, the feeding value of potatoes is amplified by cooking (which bursts the starch grains). Cooked potatoes can be preserved as potato silage. In livestock rations, the feeding value of 4 tons of fresh potatoes (for cattle) or cooked potatoes (for pigs) is roughly equivalent to that of 1 ton of maize.

Given the relatively high cost of potatoes in most developing areas, researchers and policymakers should, for the foreseeable future, concentrate on enhancing the use of the potato as a food rather than as a livestock feed or industrial input.

Price fluctuations

Price instability and price cycles. Potato prices fluctuate widely almost everywhere; however, predictable price cycles occur infrequently. Regular production and price cycles generally occur in temperate areas where there is a single, main potato harvest each year. In tropical and subtropical areas where potatoes are grown in many different ecological niches at different times of the year, prices tend to be highly unstable, and their movements tend to be unpredictable.

The economic implications of regular price *cycles* are quite different from those of the more common price *instability.* Where producers, market agents, and consumers can reliably predict the production and price cycles, they adjust their economic activities to protect them from, or take advantage of, these cycles. By doing so, they tend to moderate the cycles.

Off-season planting and storage are two examples. These activities, which are often thought to be the cause of price fluctuations, in fact tend to dampen fluctuations in market supplies and prices. Withholding supplies from the market at harvest time cushions the seasonal fall of prices, and later on, the sale of stored potatoes curbs the rise in prices in times of scarcity. Off-season planting, processing, and other strategies used by farmers and marketing agents to cope with price cycles also tend to reduce swings in prices. It is the *unpredictability* of potato production and prices in many developing countries that discourages off-season production, storage, processing, domestic transport, and international trade in potatoes. Hence, by improving estimates of future potato harvests, researchers and policymakers can help strengthen market efficiency.

Some cycles of potato prices last for years rather than months. Because potatoes cannot be stored for more than a year, even

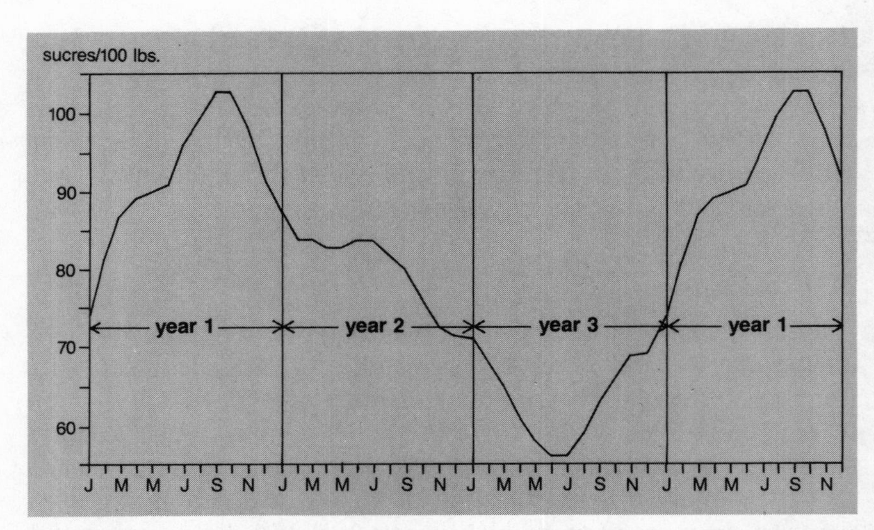

Figure 11. Ecuador: Three-year potato price cycle (3-month moving average). *Source:* Derived from M. Valderama and H. Luzuriaga, *Producción y utilización de la papa en el Ecuador* (Lima: International Potato Center, 1980).

with refrigeration, long-term cycles can discourage potato storage. Highland Ecuador, for example, has a 3-year cycle that makes potato storage profitable only 1 year in 3 (Fig. 11). In the first year of the cycle, farmers boost production in response to the previous year's high prices. The amount by which production can be increased is limited, however, by a shortage of seed tubers, which results from the low production in the previous year. In the second year of the cycle, farmers continue to expand production and prices drop substantially. This motivates producers to restrict production in the third year, causing prices to climb and triggering the cycle again.

Seasonal and long-term price cycles vary from place to place, and in some areas no cyclical patterns occur. Whenever proposals are being made to modify seeding or harvest dates or any aspect of the marketing system, local information on seasonal price movements should be examined.

Price controls. Government-enforced price controls are often proposed as a means of stabilizing potato prices. However, most price control programs for potatoes in developing countries have failed. Effective price controls require regulation of production and storage to match the quantities supplied and demanded at the desired price. This is extremely difficult with a perishable commodity like potatoes. Even in the United Kingdom, where the Potato Marketing Board strictly regulates potato plantings, yield fluctuations cause supply variations, which, in turn, cause price instability. In most developing countries, where the institutional and financial mechanisms needed to regulate production and storage are weak, it is practically impossible for a government agency to control potato prices. Most efforts to control potato prices have run into serious problems of oversupply or undersupply. In the first case, unsold stocks of potatoes have accumulated and losses have mounted until the potatoes were destroyed or fed to livestock. In the second case, an illegal black market has usually developed, with high prices for consumers and high average profits, but high risks, for traders.

Price trends and demand projections

Price trends

Long-term trends in potato production and prices can be observed in many areas. In several developing countries, potato prices have dropped as per capita potato consumption has grown. On the other hand, in several developed countries, potato prices have increased (Table 10 and Fig. 12) while consumption levels have fallen. These changes in prices and consumption levels provide important indications of the underlying movements of supply and demand (as previously shown in Figs. 7 and 8). The substantial decline in potato prices that has accompanied rising consumption in countries like India, Bangladesh, Thailand, and Colombia implies that the supply of potatoes has expanded more rapidly than demand (in economic terms, the supply curve has shifted to the right more than the demand curve). The growth in supply has resulted, in large part, from improvements in potato production and marketing technology. In contrast, demand has increased more rapidly than supply in Sri Lanka. The sharp increase in potato prices beginning in the late 1960s is one result

TABLE 10
Relative Price[a] of Potatoes in Various Locations, circa 1950 and 1980

	1950	1980
Developing countries (Potato price/rice price)		
Bangkok	3.9	0.8
Calcutta	1.6	0.9
Dhaka	2.4	0.7
Delhi	1.2	0.9
Manila	1.6	1.3
Mauritius	0.9	0.4
Seoul	0.3	0.5
Colombo	0.7	1.1
Developed countries (Potato price/wheat flour price)		
Amsterdam	0.3	0.3
Dublin	0.7	0.7
Germany (FRG), national avg.	0.2	0.5
New York	0.4	0.9
Oslo	0.5	1.3
Paris	0.2	0.4
Vienna	0.3	0.2
United Kingdom, national avg.	0.5	0.6

[a] Price ratios are 3-year averages of retail prices centered on 1950 and 1981 except for: Bangkok, 1955/57 and 1978/80; Calcutta, 1948/50; Dhaka, 1952/54; Manila, 1959/61 and 1974/76; Mauritius, 1953/55; Seoul, 1955/57.

Source: Derived from statistical tables in *International Labour Review, Statistical Annex* (Geneva, various years).

of the ban on imports of consumer potatoes since 1967. In many countries of North Africa and the Middle East, relative potato prices grew during the 1960s and 1970s partly due to subsidies on bread and other wheat products.

The ascent in potato prices as consumption has declined in European countries like Germany implies that the supply of potatoes has fallen more rapidly than the demand for them. This has been due, at least in part, to the slow pace of technological change in potato production and marketing there. (A severe drought caused the jump in potato prices in Europe in the mid-1970s.)

These findings lend empirical support to the view that consumers respond to price reductions for potatoes by increasing consumption, and vice versa. As technology is one of the key determinants of production and marketing costs, this analysis suggests a vital and often forgotten link between technological change and potato consumption trends: To the extent that research and extension programs result in lower potato prices in developing countries, greater potato consumption is likely to follow.

Demand projections

In projecting consumer demand for foods, economists commonly assume that biological needs, food habits, and relative prices are stable and do not influence trends in demand. Usually the only factors that are taken into account in demand projections are population growth, income changes, and the income elasticity of demand.

In 1971, FAO published *Agricultural commodity projections, 1970–80,* which presents demand projections for individual foods, including potatoes, for most countries of the world. More recent demand projections by the FAO do not contain estimates for potatoes specifically, but only for the broad category of starchy foods, which include cassava, sweet potatoes, yams, cocoyams, bananas, and plantains as well as potatoes. When income elasticities or demand projections for potatoes are needed in developing countries, most analysts use FAO's 1971 estimates for potatoes or the newer estimates for all starchy foods. In some cases, they borrow estimates from studies conducted in developed countries.

FAO's 1971 demand projections were based on projections for growth in population and per capita income in each country and on assumed income elasticities of demand for each food. The projected growth rates of population and per capita income came from the United Nations' standard demographic and economic models. The income elasticities of demand for potatoes were based on household surveys conducted in 17 developed and 8 developing countries, mostly in large cities, where consumption patterns, it is worth noting, differ markedly from those

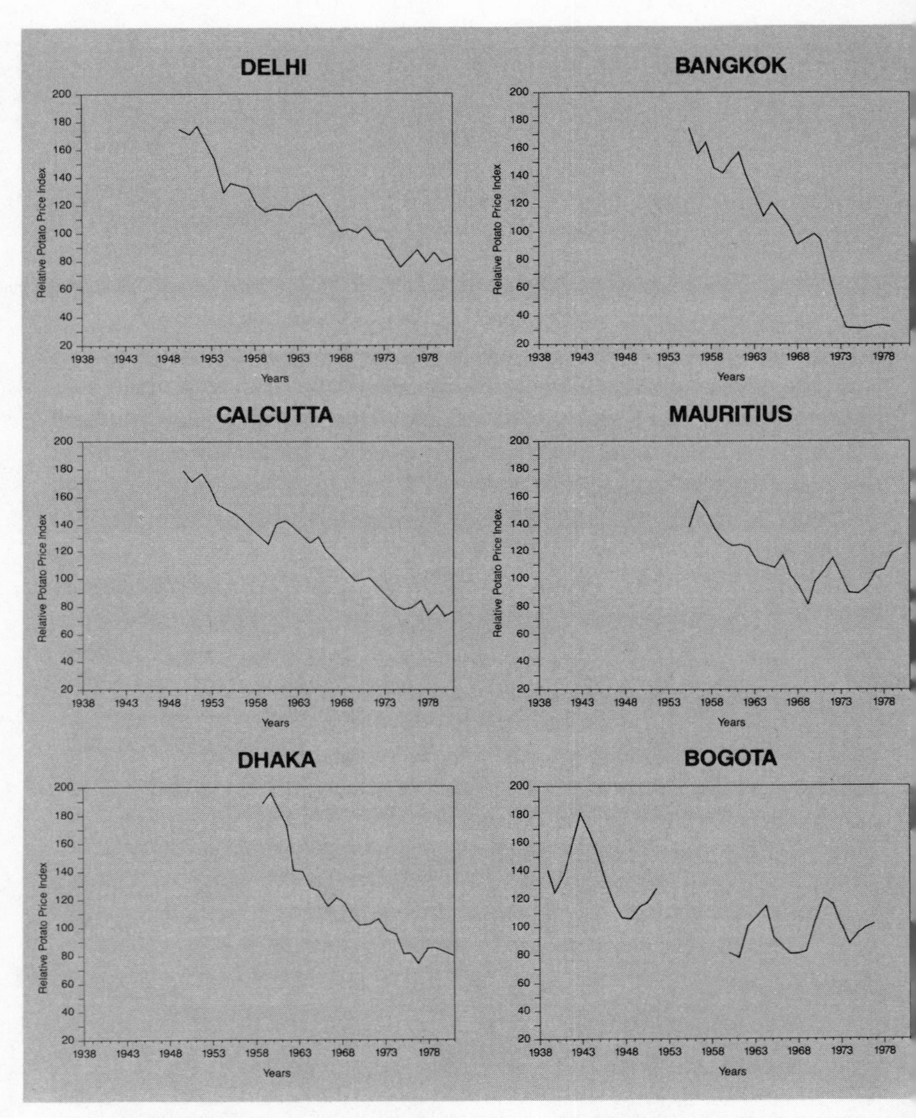

Figure 12. Trends in potato prices (3-year moving averages of retail potato prices deflated by the index of all food prices—1970 = 100). *Sources:* International Labour Office. *Statistical yearbook* (Geneva, various years); FAO, *Production yearbook* (Rome, various years).

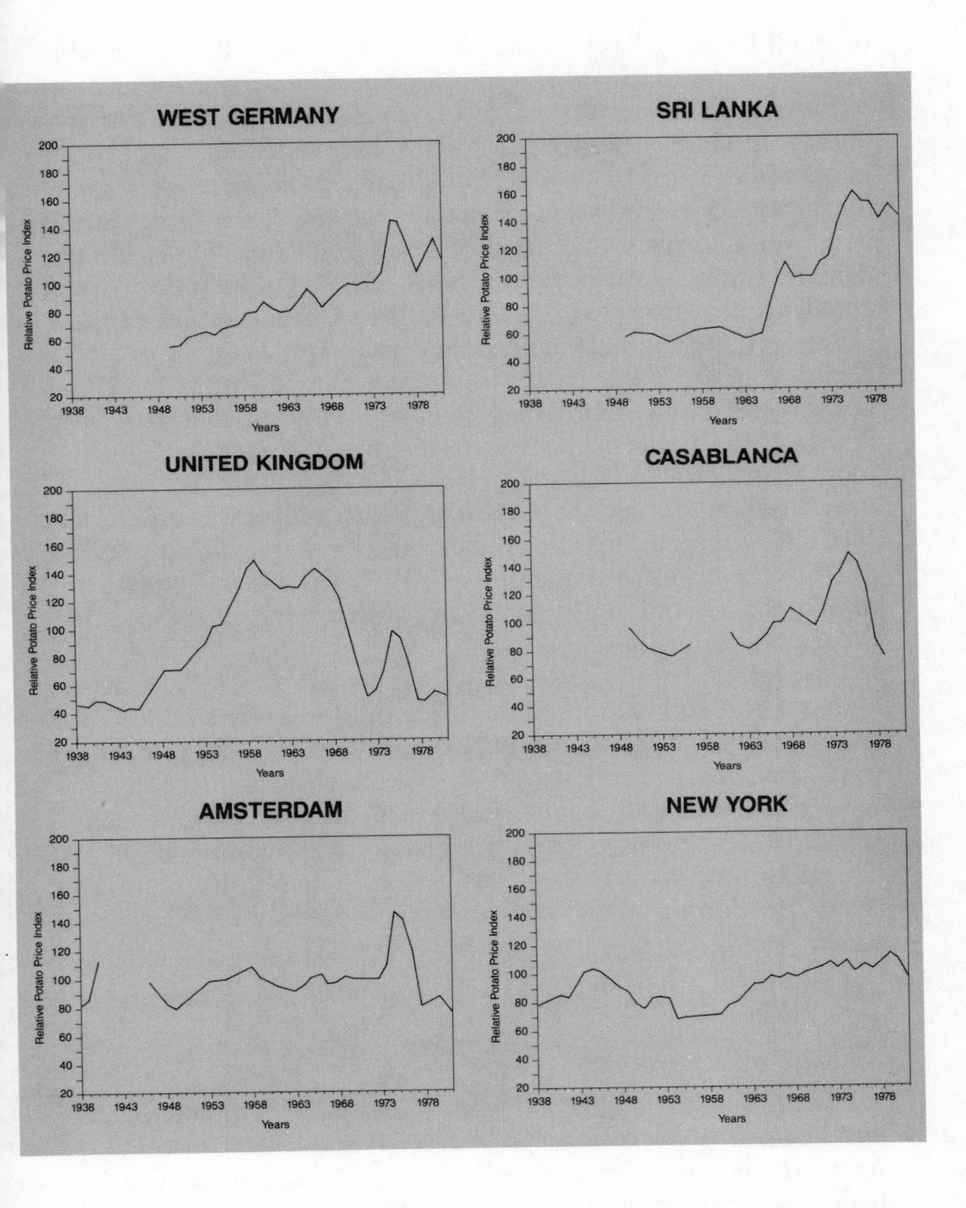
WEST GERMANY
SRI LANKA
UNITED KINGDOM
CASABLANCA
AMSTERDAM
NEW YORK
Relative Potato Price Index
Years
200
180
160
140
120
100
80
60
40
20
1938
1943
1948
1953
1958
1963
1968
1973
1978

of small towns, villages, and rural areas. In 18 of the 25 surveys, potatoes were consumed as a staple food rather than as a vegetable.

Based on the results of these few, clearly unrepresentative household surveys, demand elasticities were estimated for potatoes in all countries of the world and demand projections were made for 1980. An examination of the change that took place during that period reveals that in developed countries, the projected demand exceeded the actual demand; whereas in most developing countries, the projected demand fell short of the actual demand (Table 11). For example, consumer demand for potatoes in North Africa and the Near East was projected to advance by about 60 percent, but it more than doubled. In sub-Saharan Africa, demand was projected to rise by 54 percent, but there, too, it more than doubled. For many countries, the disparities between projected and actual changes in demand were even greater. In India, for example, growth in demand for potatoes was projected at 55 percent, but it actually was over 180 percent. Among the developing regions, only in Latin America did the projection exceed actual demand.

One reason why the FAO's projections fell short of the actual demand for potatoes in most developing countries is that the assumed income elasticities of demand were too low. But, more important, the projections did not take into account changes in food prices and food habits. Relative potato prices have tended to fall in developing countries and rise in developed countries. Changing food habits also favor greater potato consumption in many developing areas and lower consumption in developed areas.

Bibliographic notes

Chapters 1 to 4 of *Food policy analysis* (Timmer, Falcon, and Pearson, 1983) treat in greater detail and more general terms many of the concepts and principles presented here. The bibliographic notes at the end of their chapters also contain a useful selection of references on economic aspects of food consumption and nutrition, production systems, marketing, and price formation. That book is highly recommended for those concerned

TABLE 11
Comparison of FAO Projections with Actual Changes in Potato Consumption, 1965–1980

	Consumption 1965 (million tons)	Change 1965–1980 (%)	
		Actual	Projected
World	117.15	15	7
North America	11.14	22	9
Canada	1.48	29	18
USA	9.66	21	8
Western Europe	31.42	−6	1
France	5.05	−15	−3
Germany (FRG)	6.82	−27	−3
Netherlands	1.12	4	5
United Kingdom	5.54	4	6
Yugoslavia	1.25	4	17
Eastern Europe and USSR	42.05	−6	−18
Germany (GDR)	2.38	0	−19
Poland	3.99	6	−30
USSR	31.83	−9	−19
Sub-Saharan Africa	0.72	127	54
Kenya	0.11	111	82
Madagascar	0.07	71	48
Rwanda	0.03	527	65
South Africa	0.26	93	45
Zaire	0.03	342	58
Latin America	5.80	48	58
Argentina	1.47	15	27
Brazil	0.79	94	91
Chile	0.58	−12	49
Mexico	0.27	207	96
Peru	0.99	25	65
North Africa & Near East	2.08	133	63
Algeria	0.19	197	71
Egypt	0.25	234	52
Turkey	1.28	81	66

TABLE 11, cont.

	Consumption 1965 (million tons)	Change 1965–1980 (%)	
		Actual	Projected
Asia	23.23	54	42
China	16.29	55	43
India	2.29	183	55
Japan	2.64	−35	19
Korea (DPR)	0.71	58	53
Korea (ROK)	0.49	−51	47
Pakistan	0.43	−28	74

Source: Derived from FAO, *Agricultural commodity projections 1970–1980* (Rome 1971); and FAO, *Food balance sheets 1979–1981 average* (Rome, 1984). Consumption figures for China were adjusted to reflect the most recent FAO production estimates.

with any aspect of agricultural policymaking. A shortcoming of *Food policy analysis* is that, following the convention of most economics texts, it frequently cites the potato as a typical inferior good, that is, a good with negative income elasticity of demand. This is not the case in many developing areas.

The treatment of marketing issues in this chapter draws on three principal sources of information: personal experience; numerous studies conducted at the Food Research Institute, Stanford University; and studies done by Gregory Scott and others at CIP. Among the Stanford studies, a paper by W. O. Jones (1984) gives an especially insightful discussion of domestic food marketing issues and policies in developing countries and contains a useful list of references. Although Jones' article is concerned specifically with Africa and does not treat potatoes specifically, many of the conclusions are valid for other geographical areas and for potato marketing. Gray, Sorenson, and Cochrane (1954) present a fascinating discussion of the effects of immigration, World War II, price policy, and technological change on the potato industry in the United States.

Publications by Scott (1981, 1983, 1985, 1986a, 1986b) give detailed information on potato marketing in Peru, Bhutan, Thailand, and Africa. Scott's work on Peru (1985) is perhaps the most thorough study of potato marketing in any developing area. Many of Scott's conclusions are of rather general validity, and his bibliography contains numerous publications on potato marketing in other countries. Durr and Lorenzl (1980) and Durr (1980) present similar information on Kenya. Fu (1979), Valderrama and Luzuriaga (1980), and Durr (1983) discuss potato marketing in Chile, Ecuador, and Rwanda, respectively.

The most comprehensive and widely used food supply and demand projections are those of the FAO. The two-volume study, *Agricultural commodity projections, 1970–80* (FAO, 1971), provides great detail by commodity and country and outlines the assumptions and methods employed in later studies, which present estimates only for the starchy foods as a group. Due to the diverse patterns of supply and demand for potatoes, sweet potatoes, cassava, yams, cocoyams, and bananas, estimates of income elasticities and demand projections for starchy foods as a group have little value in food policy analysis or for managing research and extension programs.

R. J. Perkins, Chief of the FAO's Commodity Policy and Projections Service, has been most helpful in providing information used in preparing the section on price trends and demand projections in this chapter.

A useful and little-known source of food price data is the survey "October Enquiry," conducted by the International Labour Office in Geneva, and the results published each year in the ILO's *Statistical yearbook.* In 1924, the survey's first year, prices for 30 foodstuffs, including potatoes, were published along with representative wage rates for the United States and 13 European countries. Over the years, more countries were added to the survey, but the list of foodstuffs remained largely unchanged. Now, urban retail prices and wages are published for over 100 countries. Unfortunately, prices of many foodstuffs important in the tropics, like maize, plantains, cassava, and sorghum, are not included.

5
Potato nutrition and consumption

Nutritional aspects

The conventional view has it that the potato is a good source of energy but has little nutritional value. In fact, the potato is not an especially rich source of energy, but it contains substantial amounts of high-quality protein and essential vitamins, minerals, and trace elements (Table 12).

Composition

The nutritional value of a potato depends greatly on its dry matter content, which averages about 20 percent of the whole tuber. The nutritional value varies widely depending on variety, cultivation practices, environmental conditions, and incidence of pests and diseases. It is often stated in the scientific literature that the potato's protein content (grams of protein per 100 grams of dry matter) decreases as the percent dry matter increases. Although this is generally true, it is not wholly relevant. In the form in which they are eaten, potatoes that have a high dry-matter content also have more protein. That is, the grams protein per 100 grams fresh weight or cooked weight is positively correlated with the percent dry matter.

To evaluate potato varieties grown under different conditions, researchers should measure the nutrient composition on both a dry matter and fresh-weight basis. In addition, the nutrient levels of potatoes as eaten are important because many nutrients are lost during storage, processing, and cooking.

TABLE 12
Nutritional Composition of a Hundred-Gram Edible Portion of Various Foods

Food[a]	Water (%)	Protein (g)	Food energy (kcal)	Protein-calorie ratio (g/1000 kcal)	Fat (g)	Ash	Ca	P	Fe	Na	K	Thia-mine	Ribo-flavin	Niacin	Ascorbic acid
						(mg)									
Maize (grits)	87	1.2	51	24	0.1	0.6	1	10	0.1	205	11	0.02	0.01	0.2	0
Potatoes	80	2.1	76	27	0.1	0.9	7	53	0.6	3	407	0.09	0.04	1.5	16
Plantain	80	1.3	77	17	0.1	0.7	—	—	—	—	—	—	—	—	—
Yam (raw)	74	2.1	101	21	0.2	1.0	20	69	0.6	—	600	0.10	0.04	0.5	9
Rice	73	2.0	109	18	0.1	1.1	10	28	0.2	374	28	0.02	0.01	0.4	0
Spaghetti	72	3.4	111	31	0.4	1.2	8	50	0.4	1	61	0.01	0.01	0.3	0
Sweet potatoes	71	1.7	114	15	0.4	1.0	32	47	0.7	10	243	0.09	0.06	0.6	17
Common beans	69	7.8	118	66	0.6	1.4	50	148	2.7	7	416	0.14	0.07	0.7	0
Cassava	68	0.9	124	7	0.1	0.6	—	—	—	—	—	—	—	—	26
White bread (fresh)	36	8.7	269	32	3.2	1.9	70	87	0.7	507	85	0.09	0.08	1.2	trace

[a] Boiled unless otherwise indicated. Edible portions of potatoes and other root crops and plantains do not include peels. Dashes denote lack of reliable data.

Sources: USDA, *Composition of foods* (Washington, D.C., 1975); W-T. Wu-Leung, F. Busson, and C. Jardin, *Food composition table for use in Africa* (U.S. Department of Health, Education and Welfare, Public Health Service, Bethesda, MD , 1968).

Nutritional value

Potatoes yield about 80 kilocalories (335 joules) per 100 grams fresh weight, which is much less than the cereals (about 350 calories per 100 grams) and somewhat less than other root crops. The differences are narrowed when these foods are cooked, though tortillas and breads typically yield 200 to 300 kilocalories per 100 grams.

The potato's low energy density means it is, like boiled rice and cereals cooked as porridges, a rather bulky food. Although its bulkiness can be a disadvantage in diets for infants and children, some studies have found that potatoes can supply small children 50 to 75 percent of their total daily energy needs. The low digestibility of potato starch may preclude higher consumption levels for children. For adults, however, potatoes can supply a greater proportion of energy needs. An adult male's total daily energy requirement—2,550 calories—could be met by consuming about 3.3 kilograms of boiled potatoes. Consumption levels this high have been reported in Ireland and the Andes.

Potatoes have about 2 percent protein, a level similar to that of yams and nearly double that of sweet potatoes and cassava. Fresh potatoes have a much lower protein content than cereals, but after cooking, the potato's crude protein (total nitrogen × 6.25) is comparable to that of boiled rice and other cereals cooked as porridges. The reason is that when boiled, the potato absorbs little moisture whereas cereals absorb two to three times their weight in water.

Methionine and cystine, the sulfur-containing amino acids, are moderately limiting in potato protein. But lysine is relatively abundant in potatoes, making them a valuable addition to cereal-based diets, which are often limiting in this amino acid. The average biological value of potato protein—about 70 percent that of whole egg—is better than that of most other vegetable sources and comparable to that of cow's milk (Table 13).

The quality of potato protein is high. As little as 100 grams of boiled potato can supply 10 percent of the recommended daily allowance of protein for children and over 5 percent for adults. In feeding trials, adults have consistently maintained their nitrogen equilibrium and good health on diets in which potatoes

TABLE 13
Limiting Amino Acids and Net Protein Utilization (NPU) of Selected Proteins

	Limiting amino acid[a]	Net protein utilization[b]
Egg	—	100
Fish	Tryptophan	83
Cow's milk	S	75
Casein	S	72
Sweet potatoes	S	72
Potatoes	S	71
Sunflower seed	S	65
Rice	Lysine	57
Soy flour	Lysine	56
Millet	S	56
Maize meal	Tryptophan	55
White flour	Lysine	52
Peanut flour	S	48
Navy bean	S	47
Peas	S	44
Cassava	S	[c]

[a] "S" signifies sulfur-containing essential amino acids (methionine plus cystine)
[b] Percentage of food nitrogen retained in the body
[c] Not available

Source: World Health Organization, *Protein requirements* (Geneva, 1965).

supplied 100 percent of the nitrogen. Infants and children can safely consume as much as 80 percent of their total dietary nitrogen in the form of potatoes.

The potato is well balanced with regard to the ratio of protein to calories. The potato's protein-calorie ratio, which is about 10 percent below the protein-calorie ratios of white bread and pasta, is considerably higher than those of other root crops, most cereals, and plantains. The potato's net dietary protein calories percent (NDpCal%)—which measures a food's protein content and quality in terms of the proportion of total calories provided as protein—is near the level required by 1-year-old children and is adequate for all later age groups. Therefore, if enough potato is eaten to supply a person's energy requirement, the protein requirement will also be met.

In vitamin content, the potato is comparable to other common vegetables. One hundred grams of boiled potato provide around 10 percent of the adult recommended daily allowance for thiamine, niacin, and vitamin B6; 5 to 10 percent of the folic acid and pantothenic acid; and 50 percent of the ascorbic acid. The only staples with comparable or higher quantities of ascorbic acid are sweet potatoes, cassava, and plantain. Cereals and dry beans provide no vitamin C whatsoever, unless sprouted.

The soils in which potatoes are grown strongly influence the mineral content of the tubers. Normally, the potato is a moderately good source of iron, a good source of phosphorus and magnesium, and an excellent source of potassium.

Post-harvest influences on nutritional value

Storage, freezing, drying, and cooking tend to lower nutrient levels of all foods. In developing countries, the most significant nutrient losses in potatoes occur during storage and cooking.

Storage. Most studies of nutrient losses during storage have been conducted under controlled, low-temperature conditions. The losses that take place under the uncontrolled temperatures of on-farm storage have not been systematically studied, and that is an important area for investigation.

During cold storage, total nitrogen content is stable. Sprouting that occurs during long periods of storage raises free amino acid content slightly. Starch is converted to reducing sugars, and if the potatoes are subsequently fried, a discoloration called "browning" may lessen the appeal of french fries and potato chips; its nutritional significance is unknown.

Ascorbic acid and folic acid are the vitamins most adversely affected by storage and, unfortunately, the ones most severely depleted by cooking and processing. Ascorbic acid losses of 40 to 75 percent have been recorded after 8 months of cold storage. To minimize losses of ascorbic acid, the optimum temperature range during storage seems to be 10 to 15 degrees C. That losses increase at lower temperatures suggests that vitamin C losses may be smaller in farm storage than in cold storage. Folic acid also decreases significantly during cold storage; losses up to 40 percent have been recorded.

The level of niacin rises during the first month or two of cold storage and then declines to near that of freshly harvested tubers. Vitamin B6 increases by as much as 150 percent for reasons that are unknown. Changes in mineral constituents during storage are insignificant.

To maximize vitamin retention, stored potatoes should be steamed or boiled in their skins. Even after several months of storage, just 100 grams of potatoes that have been cooked unpeeled can still supply about 25 percent of the recommended daily allowance for vitamin C. After cooking, the levels of niacin and thiamine in potatoes that have been stored are similar to those of freshly harvested tubers, and vitamin B6 levels are higher.

Cooking. Cooking is necessary to make potatoes digestible because raw potato starch is almost totally indigestible. All cooking methods diminish vitamin content, though significant losses from intact, unpeeled tubers occur only with the heat-sensitive ascorbic and folic acids. Cooking has little effect on minerals and trace elements.

Unlike many vegetables, the potato has an edible skin that acts as a barrier to the leaching of nutrients during cooking. Thus, boiling a tuber in its skin best retains nutrients.

If tubers are peeled and sliced before boiling, nutrients are easily leached out. Peeling before cooking also removes more of the tuber with the peel and accentuates losses of minerals and folic acid, which tend to be concentrated in the peel.

Other cooking methods generally have more severe effects on nutrients than boiling, and cooking more than once compounds nutrient losses. For example, when boiled potatoes are subsequently fried, losses are much greater than when they are either boiled or fried.

Losses in tuber nitrogen resulting from cooking are usually small. Losses of all vitamins except thiamine are greater when potatoes are boiled peeled rather than unpeeled. Soaking peeled potatoes before cooking can substantially lower the content of ascorbic acid and thiamine. Holding cooked potatoes warm or chilled before eating them or preparing hash browns (boiled potatoes that are chopped and then fried), potato salads, or dumplings reduces ascorbic acid even further. Total losses of ascorbic acid can range from about 20 percent in unpeeled, boiled potatoes to nearly 90 percent in potato dumplings. For diets in which potatoes are a major source of vitamin C, they should be cooked in their skins, particularly if they have been stored.

Processing. The nutrient content of potatoes is strongly affected by many processing operations. Processing may accentuate the nutrient reductions that took place in storage. When processed products are later cooked and served, the content of vitamins—particularly ascorbic and folic acids—may fall to negligible levels.

Dehydrated products tend to have low nutrient contents. Potato chips suffer large losses of amino acids and vitamin C. However, reduction of the moisture content to about 2 percent concentrates nutrients, and chips are a reasonably good source of most nutrients.

Toxic constituents

There is no evidence that high levels of potato consumption are associated with any nutritional disease. Although the potato contains three potentially toxic substances—glycoalkaloids, proteinase inhibitors, and lectins—few instances of fatalities from toxicity related to heavy potato consumption have been recorded.

Glycoalkaloids occur mostly in the outer layers of the tuber, so peeling tends to remove them. Preventing prolonged exposure of tubers to light during harvest and subsequent handling retards accumulation of glycoalkaloids. Because the glycoalkaloid levels of varieties differ, and they play a role in disease and pest resistance, it is important for potato breeders to screen potential new varieties for glycoalkaloid content.

Proteinase inhibitors, which can constitute more than 15 percent of the mature tuber's soluble protein, are destroyed during cooking, and no adverse effects on humans have been reported. The nutritional significance of potato lectins is not well understood. Because they too are destroyed by heat, no ill effects are likely to result from cooked potatoes, and no link has ever been reported between potato consumption and lectin toxicity.

Potato consumption

Since its domestication in the Andean highlands of South America, the potato has spread throughout the world. It has adapted not only to a variety of growing environments, but it also has found a place in diverse human diets. To some people, the potato is a delicacy; to some, it is a staple food; and to others, it is a taboo item, to be avoided or eaten with caution. When planning and implementing agricultural programs, researchers and policymakers usually focus their attention on expanding production. Higher food production is, however, not sufficient to ensure greater consumption and improved nutrition. The potential for raising consumption of a food is influenced by its role in the diet and the extent to which that role is likely to be altered in response to changes in supply. Hence, in planning and implementing a potato program it is essential to consider not only production, storage, and marketing, but consumer behavior as well.

Estimating potato consumption

Estimates of potato consumption are usually based on food balance sheets. As shown in Table 4 in Chapter 1, the food balance sheet equates domestic production plus imports minus exports and net changes in year-end stocks to total domestic availability—the sum of the quantities used for seed, industry, animal feed, waste, and human consumption. Human consumption is, in fact, generally estimated as the residual. It is then divided by the estimated national population to obtain per capita consumption.

Unfortunately, the information used in the equation is often unreliable. On the supply side, determining total production is difficult because potatoes are often grown in isolated mountainous areas on small irregular plots, frequently as intercrops, relay crops, secondary crops, or backyard garden crops. Consequently, potatoes may easily be overlooked in national statistics.

On the utilization side, the figures are equally suspect. Seeding rates in the developing countries are usually assumed to be close to rates used in industrial countries. However, because farmers in poor countries tend to use small tubers for seed, the tonnage of seed planted may be vastly overestimated. Compilers of food balance sheets may indiscriminately apply waste factors of 10, 15, or 20 percent without regard to indigenous practices in peeling, cooking, or utilization of peels. So far as is known, no developing country's estimates of seed, waste, or processing are based on local, on-farm, or market-level research. Instead, they are guesses.

Aside from the problem of data accuracy, food balance sheets have the limitation of all average national estimates. Their figures do not reflect important differences in consumption that occur between regions and across social groups. Yet despite these limitations, food balance sheets provide a general indication of consumption levels and trends and serve as a starting point for estimating potato consumption. In addition, because most official statistics are based on food balance sheets, it is advisable to be familiar with this data.

Household surveys provide an alternative for estimating potato consumption. Well-planned surveys can uncover seasonal patterns of potato consumption as well as differences between regions and social groups. Results of household surveys conducted in many countries imply that average consumption levels are considerably higher than indicated by estimates derived from food balance sheets.

Tunisia provides an example of the discrepancy between results of food balance sheet calculations and household surveys. Based on the Ministry of Agriculture's official production estimate, the food balance sheet puts potato consumption at about 18 kilograms per capita. Recent nationwide household consumption surveys, however, estimate average potato consumption to be closer to 25 kilograms.

Unfortunately, surveys usually are done in urban areas; figures for rural areas are rarely available. Another limitation of published surveys is that few report specific consumption estimates for potatoes. Instead, they give estimates for all root crops or starchy foods in general. For these reasons, special studies are needed to determine potato consumption patterns in many areas.

Consumer groups

In most countries, there are substantial differences in the potato consumption patterns of potato producers, urban dwellers in potato-producing areas, residents of major consumption centers, farmers in nonproducing zones, expatriates and minority ethnic groups, and institutional consumers.

The main market flow of potatoes is from producing areas to major urban markets. From the latter, only a small proportion of marketed potatoes may move to village distribution centers and rural areas. Because of the accumulated costs of marketing potatoes, the retail price is generally lowest in rural producing areas and progressively higher in towns in rural producing areas, in regional and national capitals, in towns in nonproducing areas, and in rural nonproducing areas. Consumption levels tend to vary inversely with price levels. Moreover, within each area, potato consumption will normally vary with income level. In developing countries, excluding Andean and temperate South America, wealthy people usually eat more potatoes than poorer people in the same locality.

Potato producers have better access to cheap potatoes than does any other consumer group, and hence, they tend to consume more than other groups. If potatoes are an expensive vegetable, however, wealthier urban consumers and foreigners may have the highest consumption levels.

Aside from growers, potato wholesalers and retailers have the most ready access to inexpensive potatoes. Market centers in producing zones have lower potato prices than other cities, especially at harvest times. Therefore their consumption levels can provide an indication of the potential urban demand for potatoes at somewhat reduced prices. Farm families outside potato-producing zones have the poorest market access to potatoes and must pay the highest prices for them. They also tend to have low incomes and ready access to cheaper food they produce themselves. Consequently, these consumers usually eat fewer potatoes than other groups do.

Every country has groups whose consumption patterns differ from those of the general population because of special economic, social, cultural, or political attributes. Foreigners, for example, bring with them a set of food habits that never become completely accommodated to the new environment. In most developing countries, foreigners from Western Europe tend to consume more potatoes than does the native population. Different ethnic groups sharing the same place of residence, having the same income level, and facing the same prices may also have markedly different levels of potato consumption. In Indonesian cities, for example, the ethnic Chinese eat more potatoes than do non-Chinese city residents of the same socioeconomic level.

Surveys conducted in countries such as Indonesia, Guatemala, and Rwanda indicate that within a locality, whether urban or rural, per capita potato consumption is positively correlated with household income. Between areas, consumption was found to be highest in potato-producing zones and in large cities, and lowest in nonproducing rural areas. In potato-producing areas, wealthier farm families consumed more potatoes than did the poorer ones. In some instances, this pattern relates to the lack of storage facilities of poor farmers, but usually it stems from the tendency of poor families to sell a larger portion of their crop and to purchase cheaper foods.

Surveys in Peru reveal that consumption levels differ sharply between the country's three major agroecological regions. In the highlands, potatoes are a major staple food: In many areas per capita consumption is around 0.5 kilogram per day, or over 150 kilograms per year. On the coast, where other low-cost staple foods, such as noodles, bread, and rice compete with potatoes, urban consumption is 50 to 100 kilograms per year and average rural consumption is lower and highly seasonal. In the Amazon basin, where potatoes cannot be grown, per capita potato consumption is very low. Potatoes are costly to transport into the Amazon region and storage is extremely difficult in the warm, humid climate. As a result, people rely on local foods such as rice, plantains, and cassava and to a lesser extent on imported cereals and pulses.

Consumption patterns

The examples cited above illustrate the great diversity that exists in potato consumption practices. Nevertheless, certain patterns can be detected in the frequency and quantity of potatoes consumed and the relationship of potatoes to other food items.

Potatoes can occupy one of four positions in a meal. They may be the staple food, as in much of Europe in the nineteenth century, and in certain developing areas today, such as highland Rwanda, India, Nepal, the Andes, and temperate zones of South America and China. In such areas, spices and sauces add variety to the ubiquitous plate of potatoes. Second, the potato may be one of several staple foods. In Peru's city markets, for example, potatoes, rice, and noodles may all be served on the same plate in roughly equal portions. Third, the potato may be an important side dish served along with one or more staple foods. In this case, potatoes complement the meal, but are not a major source of energy. This pattern is found in Central America where maize tortillas and beans are the dietary staples, and potatoes are served regularly but in small quantities as a vegetable. Likewise in India and Bangladesh, potatoes are frequently eaten as a vegetable, complementing rice and wheat, the staple foods. In these countries, potatoes are now considered "the poor man's vegetable," the one that is cheapest and most widely consumed by all sectors of society. Finally, potatoes may be only one of several side dishes that complement the basic staples. This pattern is common in Indonesia, Malaysia, the Philippines, and other Asian countries where potatoes are more expensive than most other vegetables.

Certain beliefs concerning potatoes are linked to their role in the diet. Where potatoes are a staple food, they are usually valued for their energy-producing qualities and are grouped with grains or other root crops rather than with vegetables. In contrast, where potatoes are consumed as a delicacy, they are as a rule not viewed as having much food or energy value. In many Asian countries, food and rice are synonymous, and all other items,

Throughout Asia, potatoes are marketed side by side with other fresh vegetables.

including potatoes, are merely trimmings, garnishes, or flavorings. In this role, potatoes are grouped with cool-season, "luxury" vegetables such as cabbage, cauliflower, carrots, cucumbers, and celery. In the Baguio city market in the Philippines, potatoes are sold in plastic bags, as evidence of their high status.

Some groups of people attach special social significance to potatoes. In Indonesia, for example, potatoes are exchanged in lavish dishes to cement social ties during celebrations following Ramadan, the Moslem month of fasting. In the Andes, peasant farmers exchange prized native potatoes called *papas de regalo* (gift potatoes) among relatives, close friends, and ritual kin.

Factors influencing consumption

Several factors influence the quantity and frequency of potato consumption. The importance of price and income level has already been treated in Chapter 4. In terms of food energy and

In the Baguio city market in the Philippines, packaging potatoes in plastic bags indicates their status as a luxury food.

protein, potatoes are relatively more expensive in the tropics and subtropics than in Europe (Table 14). Because incomes are lower in developing areas and potato prices are higher, most people in these areas cannot afford to eat as many potatoes as people in Europe do.

TABLE 14
Average Potato Price in Relation to Prices of Wheat Flour, Rice, and Urban Wages, Major World Regions, 1980

Region	Retail price[a] (US$/t)			Relative potato price		1980 brick-layer's wage (US$/hr)	Potato purchasing power[b] (kg/hr)
	Potato	Wheat	Rice	vs. wheat flour	vs. rice		
Western Europe[c]	366	719	1,437	0.51	0.25	6.35	17
Sub-Saharan Africa[d]	720	806	883	0.89	0.82	0.82	1
N. Africa & Near East[e]	390	369	651	1.05	0.60	2.58	7
Latin America[f]	560	643	931	0.87	0.60	1.54	3
Asia[g]	435	511	528	0.85	0.82	0.49	1

[a] Converted from local currency with official exchange rate
[b] Wage rate divided by potato price
[c] 13 countries
[d] 7 countries
[e] 6 countries
[f] 14 countries
[g] 10 countries

Source: Derived from International Labour Office, *Bulletin of Labour Statistics, 1981-82* (Geneva, 1983).

Taboos, famines, foreign influences, and social status can also shape potato consumption patterns. When potatoes were introduced into Rwanda, they were considered taboo items and were not consumed at all by local people. If one ate them, it was believed that one's cows would become sick and die, or the cows' milk would go bad.

European colonists greatly influenced eating habits in developing countries. Colonial administrators often introduced potato cultivation. In Rwanda, Belgian schoolmasters and missionaries encouraged boarding school children to eat potatoes. Dutch varieties introduced into Indonesia are still widely grown and preferred for their white skin and yellow flesh. The English promoted consumption of potatoes in Kenya as well as throughout South Asia. In the Philippines, consumption grew after the American occupation.

In Southeast Asia, although potatoes are expensive relative to staple foods and indigenous vegetables, they are also associated with prestigious western eating habits. Thus, conspicuous consumption of potatoes is a way of demonstrating affluence and assimilation of western culture. In Singapore, eating McDonald's french fries is a status symbol among teenagers.

Type and quality of potatoes

Consumer preferences for types and qualities of potatoes vary greatly among locations, and even within a location, consumers may prefer different potatoes for different food preparations. In the Philippines, for example, cooks in Luzon choose red-skinned tubers for salads and white-skinned potatoes for cooking with meat or vegetables. But in Mindanao, consumers prefer white-skinned potatoes over red-skinned potatoes, which are believed to spoil quickly. In Indonesia, red- or purple-skinned potatoes are rejected by consumers because they look like sweet potatoes, a low-status food.

In many countries, consumers prefer potatoes grown in certain areas. For example, in Guatemala, the village of Santa Rosa was for many years a leading potato-producing area, and its potatoes were highly esteemed by urban consumers. By the mid-1980s Santa Rosa's potato production had fallen sharply; nevertheless, merchants sell potatoes from many locations as "Santa Rosa potatoes," and consumers continue to pay top prices for them.

The significance attached to tuber size and shape, skin and flesh color, and cooking quality depends in part on how the potatoes will be used and also on the consumer group involved. Most consumers want large potatoes rather than small ones, but in many places—ranging from Peru and Colombia to Nepal, Bangladesh, and Indonesia—small potatoes of traditional varieties are highly prized for certain dishes. Moreover, medium-sized potatoes often sell at higher prices than the largest ones, which often are hollow or cracked. In some areas—such as highland Peru—consumers prefer a potato that has a high dry-matter content because it holds its shape when boiled; whereas in other areas, a potato with low dry-matter content is preferred because it disintegrates when boiled—as in Chile—or because it absorbs little oil when fried—as along Peru's coast.

The broadest generalization that can be made is that no matter where or in what role potatoes are eaten, consumers seek and pay premium prices for potatoes with specific culinary qualities that are, in turn, associated with varietal names; sizes and shapes; distinctive colors; and in many cases, production zones. Any new potato variety, or potatoes from a new production zone, must not only produce well, they must meet the requirements of local consumers or be penalized in price.

Bibliographic notes

Many studies have been conducted on the nutritional value of the potato. Unfortunately, most publications are narrow in focus, highly technical, and inaccessible to nonspecialists. Numerous criteria are used for nutritional evaluation, and materials and methods differ between studies. Consequently, many apparently contradictory findings result from different authors measuring different things or measuring in different ways.

In contrast to the vast literature on the nutritional value of the potato, little has been written on potato consumption, especially in developing areas. In order to make reliable information on potato consumption and nutrition available to researchers and policymakers, CIP sponsored a series of potato consumption studies in developing countries, some of which are reported in Poats (1983), and arranged for preparation of a book detailing nutritional aspects of the potato (Woolfe, 1987).

The section here on nutritional aspects is based entirely on Woolfe (1987). The section on potato consumption is based on Poats (1983).

6

Potato production systems

More has been written on how potatoes *should* be grown in developing countries than on how they actually *are* grown. Research publications that deal with potatoes report mostly on specialized technical studies conducted in laboratories or on experimental plots. Inadequate information on how farmers in developing countries produce potatoes and on what obstacles exist to expansion of production and use has hampered many potato research and extension programs. To help overcome these problems, this chapter provides information on actual farming practices.

Potatoes are grown from sea level to over 4,000 meters elevation, and from the equator to more than 40 degrees north and south. This vast area contains an immense array of agroclimatic conditions. Each local environment presents a specific set of opportunities for and constraints to potato production, which are reflected in farmers' practices. Most environmental variables that influence potato production, such as elevation and latitude, are continuous rather than discrete. As it is impossible to describe a continuum, it is convenient to think in terms of a simple typology of major potato growing zones in developing countries. For the purposes of this discussion, four major zones are considered: highland tropical and subtropical zones, lowland tropical and subtropical zones, temperate zones, and Mediterranean zones (Fig. 13).

Tropical and subtropical zones can be defined as areas in which average monthly temperatures are 10°C or higher. Mediterranean zones can be defined as areas that have a dry summer and a humid, low-sun period in which the coldest month has

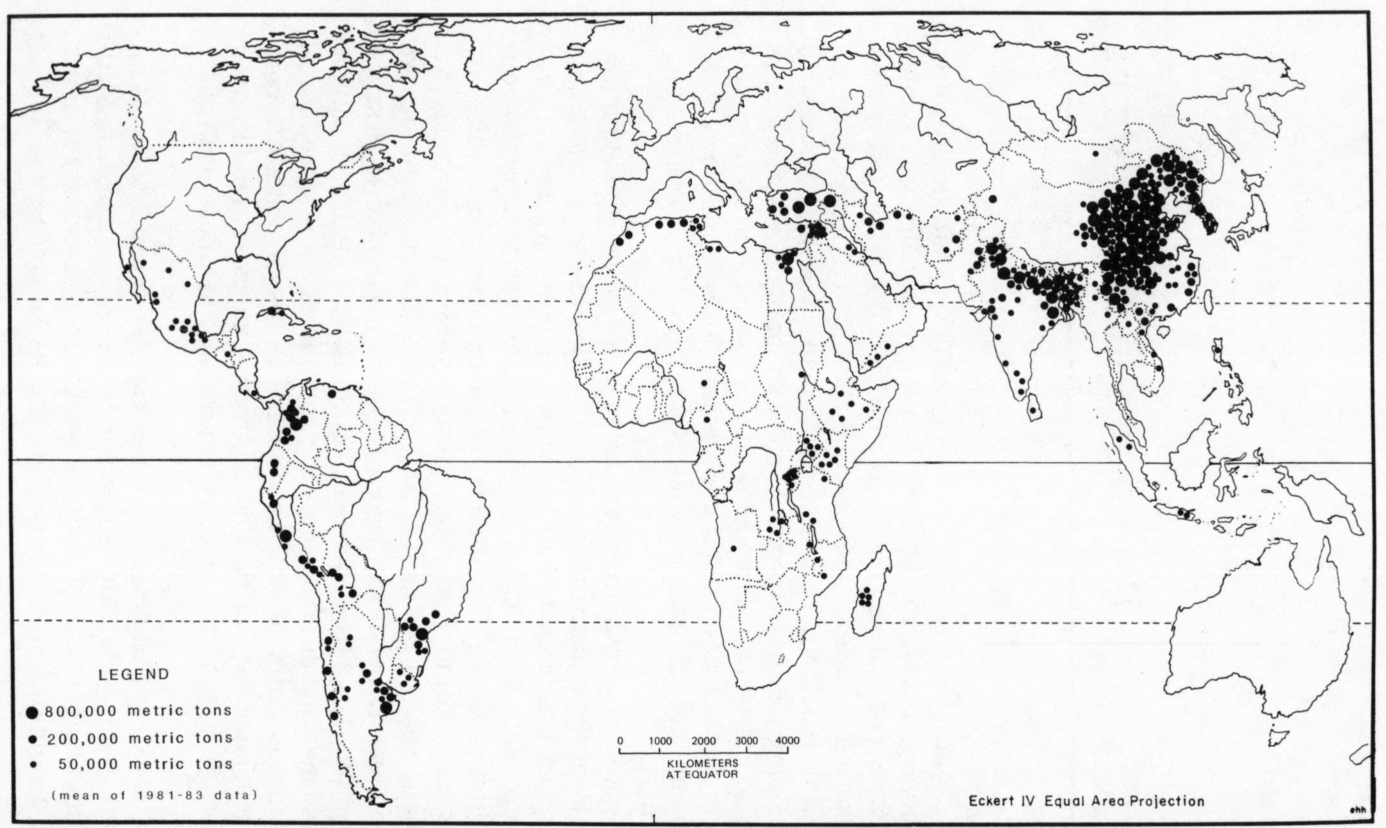

Figure 13. Approximate location of major potato-producing regions in developing countries (production not shown for countries producing less than 25,000 tons). *Source:* R. Rhoades, *Potato production zones and systems of developing countries* (Lima: International Potato Center, in press). Cartography by Robert and Ellie Huke.

an average temperature between 0 and 10°C. Temperate areas have at least 1 month with an average temperature lower than 0°C. Under this classification scheme, temperate zones include a wide range of climates, primarily in China, which are more diverse than those typical of northern Europe and North America.

Major tropical and subtropical highland potato-producing zones are in the Andes of South America, southern Brazil, Mexico, Central America, central Africa, and the Himalayas. Virtually all of the potatoes produced in Bolivia, Colombia, Ecuador, Ethiopia, Kenya, Madagascar, Malawi, Nepal, Rwanda, Uganda, Yemen, and Zaire are grown in highland areas. In contrast, all the potatoes produced in Bangladesh, and most in Burma, Cuba, India, and Pakistan are grown in lowland areas.

Potato-growing areas in China, Bhutan, North and South Korea, Afghanistan, and Argentina are mostly in temperate zones. Many of these areas are in the highlands. Using Koppen's climatological classification, potatoes in China are cultivated in at least eight different climatic regimes that differ according to their summer and winter temperatures and rainfall. Hence, within the broad temperate category, growing environments range from hot, dry summers and cool-to-cold, dry winters (northern China) to warm-to-hot, wet summers and cool, dry winters (southern China). As China's potato production exceeds that of all other Asian, African, and Latin American countries combined, the temperate zone accounts for a substantial portion of the potatoes grown in all developing countries.

Mediterranean areas account for a small fraction of the developing world's potatoes; nevertheless, important producing regions are located in Algeria, Morocco, Tunisia, Cyprus, Iran, Iraq, Chile, and Turkey.

Cultivation methods and post-harvest practices in these zones differ. Production systems in temperate and high elevation areas, particularly over 2,000 meters, have more in common with each other than with lowland production systems. Systems in Mediterranean and mid-elevation highland areas are intermediate in many respects. For example, temperate and high elevation areas have cold winters, so farmers usually grow potatoes in the summer. Lowland zones have hot summers, so farmers usually grow potatoes in the winter. In mid-elevation and Mediterranean zones, farmers grow potatoes at various times of the year, depending on local climate and market conditions. Seed potatoes are usually produced in temperate and highland zones and shipped to lowland and Mediterranean zones. Storage is easier in temperate and highland zones, where potatoes are harvested before the cool season, than in lowland tropical zones, where potatoes are harvested at the beginning of the hot season.

Within each broadly defined zone, production systems vary, depending upon local ecological and socioeconomic conditions. Potatoes may be grown as a subsistence crop for household consumption or as a cash crop for sale. As a cash crop, potatoes may be produced in field crop systems or in more intensive market-gardens.

In subsistence areas, farmers seldom plant more than a quarter hectare of potatoes, because planting that small amount of land can produce over a ton of potatoes, which is more than most families, even in the Andes, consume in a year. Yields of subsistence growers are generally low because no purchased inputs are used. Nevertheless, they often produce potatoes quite efficiently in terms of the value of output generated with the resources available to them.

In commercial field-crop areas, both large and small potato farms may be found. Typically, small farms are numerous; the farmers grow potatoes for home consumption, purchasing few inputs and harvesting low yields. The few large commercial farms tend to employ more advanced production techniques and achieve higher yields. Few farmers specialize in potatoes because of the high costs and risks involved. Most employ farming systems involving several crops and livestock. Diversification helps farmers to spread input demands through the growing season and to reduce risk. Rotations and mixed cropping help suppress the buildup of diseases and insects.

In market-gardening areas, where all potato production is intended for sale, land is usually costly, farms tend to be small, input use is intensive, and yields are high. Generally, even small farmers are relatively prosperous due to the ample yields, the possibility for multiple cropping (often three or more crops per year), and the high value of potatoes and other vegetables.

Household gardens are the least known of all potato production systems. Agricultural research and extension programs usually focus on medium-sized and large farming enterprises; household gardening is usually considered to have little economic or nutritional significance. In fact, the household garden, the most ancient and persevering form of cultivation, is more important in food production than is commonly recognized.

Household gardens make intensive use of household management and labor. In many parts of the world, they have played key roles in plant migration and agricultural experimentation. Due to the sexual division of labor within the family, household gardens are often the responsibility of women, whose highest priority is usually on meeting family food needs. Thus, improvements in household gardening can have a strong direct effect on family nutrition. As a secondary goal, individual families may produce vegetables for sale in local markets. In Central America, Africa, and Asia, potatoes are one of the favored highland garden vegetables.

Highland potato production

Potatoes are grown at higher elevations than any other major food crop. Farmers in the Andes plant potatoes in fields as high as 4,300 meters above sea level. In the tropics, the typical mountain area that produces potatoes is cool, but temperatures fluctuate sharply from day to night, and the average relative humidity is high. Soils are usually well drained, but there is great variation in altitude, slope, soil fertility, and other environmental variables that influence yields. As a result, potato production methods and yield levels may differ greatly from field to field and from season to season. Production hazards like hail and frost cause low average yields in many highland areas. Where the chances of crop failure are great, farmers often economize on purchased inputs in order to minimize their financial risks.

Because of their remoteness, many mountainous areas have lagged behind lowland areas in agricultural development and market integration. However, where road construction into mountainous areas has reduced transport costs, the introduction of

In Almolonga, Guatemala, potatoes are an important constituent of high-yielding vegetable-cropping systems.

new crop varieties and production methods has stimulated commercial production of vegetables, including potatoes, for sale in lowland urban markets.

Producing areas. In the Andes and in central Africa, potatoes are planted along with other tubers, grains, and pulses in mountainous areas above 2,000 meters. Representative high areas where potatoes are grown include northern Argentina, the altiplano of Bolivia, the Páramos of Colombia, central and southern Peru, and the highlands of Ethiopia, Kenya, and northern Rwanda.

Potatoes are cultivated in mid-elevation areas scattered throughout Central and South America; the Far East; and Africa, south of the Sahara. Representative mid-elevation zones include southern Brazil; most of the Colombian and Ecuadorian highlands; the northern and eastern slopes of the Peruvian Andes; Mexico's central and southern highlands; northern India around Simla; most of Nepal; the Dalat area of Vietnam; and mountainous

In mountainous areas of tropical Asia, potatoes are grown on paddy land in rice-based cropping systems.

areas on many islands, like the Dominican Republic, Haiti, Jamaica, the Philippines, and Sri Lanka.

Cropping patterns. At very high altitudes in the Andes and the Himalayas, where harsh climate precludes cultivation of many food crops, potatoes are often the dominant crop and compete with livestock as the principal agricultural activity. At somewhat lower elevations, potatoes are one of several crops grown during summer. And at yet lower elevations in the mountains, potatoes compete with a large number of crops for a place in highly varied year-round cropping systems.

In the highest areas, where farmers can grow only a few hardy crops, such as potatoes and barley, their typical crop rotations include fallow. There are numerous reasons for fallowing. Where population pressure is low, fallowing is an economical way of restoring soil fertility. The alternative—application of fertilizer—

would raise production costs and magnify the economic loss if the crop failed. Fallowing also reduces populations of soil-borne pests that attack the potato crop. In South America, the period between potato crops in the traditional 7-year Incan rotation was just sufficient to reduce nematode populations to economically insignificant levels. Another reason for fallowing is to provide pasture for livestock, which is a major farming enterprise in many highland areas. In rotations that include fallow, potatoes are normally the first, and best-fertilized, crop in the rotation. Usually potatoes are grown in rows; mixed cropping is unusual.

At intermediate elevations, where potatoes are grown in rotation with other crops, row cropping and sole cropping are the norm, but interplanting is sometimes practiced. Fallowing is uncommon. In field crop areas like northern Peru and northern India, potatoes are typically rotated with cereals and pulses. In market-garden areas, like Benguet in the Philippines and Badulla in Sri Lanka, the potatoes are generally rotated with vegetables.

At low elevations in mountainous areas, farmers have created more complex and diverse cropping systems. Mixed cropping is more prevalent, and beds with complex spatial arrangements of different crop species sometimes replace row cropping. In areas with natural forests, potatoes are sometimes planted as a component of slash-and-burn (swidden) farming systems, along with maize, beans, cassava, sweet potatoes, and other food crops.

In high areas where planting dates are dictated by climatic cycles, early and late plantings may be possible in irrigated, frost-free areas. Farmers in lower areas, where frost is not a problem, have more choice in their planting decisions. Nevertheless, the threat of typhoons or outbreaks of late blight may discourage potato cultivation in certain seasons.

Varieties. The variety a farmer grows often reflects historical patterns of colonization and trade, as well as the production environment, cropping system, food requirements, and consumer preference. Native varieties of subspecies *S. tuberosum andigena* are cultivated throughout the Andes of South America but nowhere else. In areas that have a long growing season (5 to 7 months) followed by an equally long cool period, farmers prefer late-maturing varieties that have a long period of dormancy. These characteristics allow farmers to take full advantage of the growing season and to store potatoes for later consumption, sale, or use as seed. In the high Andes, where frost is unavoidable, farmers grow frost-resistant bitter varieties, which must be processed before they can be eaten. For other potato growers, frost is infrequently a problem, and they plant varieties that can be consumed without processing.

Farmers in areas where late blight is common seek resistant varieties. If environmental conditions permit cultivation of other crops before and after the potato (for example, in most vegetable-growing areas), farmers usually select fast-maturing varieties. In areas where seed tubers must be held for several months before planting, varieties that have long periods of dormancy are preferred; where seed tubers need to be stored for only a short period, short-dormancy varieties, which can be planted soon after harvest, are preferred.

In selecting a variety, farmers consider cooking characteristics, eating quality, and market preferences in addition to yield and storability. Consumer tastes vary widely. Consumers generally prefer, and pay premium prices for, familiar local varieties. For this reason, in some areas it is profitable for farmers to grow low-yielding older varieties that fetch a high market price.

Seed systems. In the tropics and subtropics, most potato growers are aware that seed tubers from high areas—environments unfavorable for virus-transmitting aphids—are usually healthier and produce more vigorous crops than tubers grown at lower elevations. As a result, in many countries intricate informal seed systems have developed that link seed potato producers in high areas with growers in lower areas. Highland farmers, who usually lack irrigation, often plant smaller seed tubers and less seed per hectare than extension agents recommend. As average potato yields are low and crop failure is frequent, this practice reduces production risk. Farmers in lower areas, where irrigation is more common and yields are higher and more stable, generally plant larger seed tubers and more seed per hectare.

Cultivation. Potatoes require good soil preparation for high yields. In mountainous areas, because it is often difficult to break sod with draft animals after fallow, farmers use hoes or specialized implements, such as the Andean footplow, though in a few areas tractors are used. In market-gardens, due to the small size of plots and the high value of the potato crop, most operations are done by hand.

As a rule, highland farmers fertilize potatoes more heavily than their other crops. Use of barnyard manure is common. If manure is not available locally, farmers haul it in from livestock-producing districts. Chemical fertilizer is now widely used in most potato-growing areas. Farmers tend to use higher rates on irrigated crops than on rainfed crops and in market-gardens than in field-crop production systems. But farmers at high elevations, where the risk of crop failure is great, rarely use much chemical fertilizer. And in remote areas such as northern Rwanda and eastern Guatemala, where chemical fertilizers are unavailable, potatoes are grown without them. Farmers apply organic matter, and as soil fertility declines, they shift cultivation to new areas.

The potato has fewer pest problems in highland areas than in lowland areas. Nevertheless, as cropping intensifies in mountainous regions, the incidence of pests rises, too. Use of pesticides is expanding as chemicals and sprayers become available. In some areas, however, lack of water for mixing chemicals in the field prevents farmers from spraying.

Late blight is usually the most serious disease problem in highland areas, and bacterial wilt is an important production constraint in many mid-elevation zones.

As potato farmers cannot cure diseases that attack their crop, they employ a number of strategies to avoid them. To prevent attacks of late blight, more farmers plant resistant varieties if they are obtainable. They may also spray if fungicides are readily available. To minimize bacterial wilt, farmers avoid planting potatoes in fields where the disease is known to be present. Many farmers also "live with" bacterial wilt by rotating potatoes with wilt-resistant crops. Farmers attempt to avoid attacks of virus diseases by acquiring their planting material from higher areas that are relatively virus free.

Production costs. The level and structure of production costs differ greatly among locations (Table 15). Costs are generally over US$1,000 per hectare in commercially oriented systems. They are highest in market-gardening areas like the Benguet

TABLE 15
Distribution of Expenditures on Inputs for Potato Production in Various Areas

Area	Cost (US$/ha)	Distribution (%)					
		Seed	Labor	Fertilizers	Pesticides	Power, equip., fuel[a]	Total[b]
				Highland Areas			
Ruhengeri, Rwanda	300	38	62	0	0	0	100
Kenya	600	21	54	12	7	6	100
Sabana de Bogotá, Colombia	1200	24	34	22	14	6	100
Benguet, Philippines	1900	55	16	17	10	2	100
Quezaltenango, Guatemala	1500	53	10	17	15	5	100
				Lowland Areas			
Cañete, Peru	1700	38	20	17	11	14	100
Dhaka, Bangladesh	2600	37	27	20	15	1	100
Punjab, India	n.a.	45	27	18	6	4	100
				Temperate Areas			
Puerto Varas, Chile	2400	33	15	21	31	0	100
South Korea	n.a.	32	38	3	26	1	100

[a] Includes draft animals, tractors, tools, and implements
[b] May not add to 100 due to rounding

Sources: D.E. van der Zaag and D. Horton. Potato production and utilization in world perspective with special reference to the tropics and sub-tropics, in *Research for the potato in the year 2000,* ed. W. J. Hooker, 44–58 (Lima: International Potato Center, 1983); and D. Horton et al., Root and tuber crops in developing countries, in *Proceedings of the sixth symposium of the International Society for Tropical Root Crops* (Lima: International Potato Center, 1984).

province in the Philippines, and lowest in subsistence-oriented areas like northern Rwanda. Seed tubers account for as little as 20 percent of the production cost in areas like Kenya and the Sabana de Bogotá in Colombia, where reasonably good quality seed is produced locally. In other areas, such as Guatemala, where appropriate sites for seed production are limited, or the Philippines, where costly seed tubers have to be imported, seed may account for over half the total production cost. Use of chemical fertilizers and pesticides depends on availability and price. Farmers in Rwanda do not use chemical inputs at all because they are not available at local markets. In other highland areas, most of the pesticide cost is for fungicides. The heavy expense of imported seed in places like the Philippines underscores the need to develop cost-effective systems for multiplying imported seed or producing high-quality seed locally.

Harvesting and post-harvest technology. Potatoes can be harvested at any time after tuber formation begins. The option of early harvesting—that is, prior to senescence—allows potatoes to fill niches in many cropping systems that are too short to permit the growth of a cereal or pulse crop with their fixed and rather long growth periods. Despite the small yield of immature tubers, and their vulnerability to damage in handling and storage, farmers often decide to harvest potatoes early to capitalize on seasonally high prices. "Early" potatoes usually sell for two or more times the price of main-crop potatoes. Another reason farmers may harvest early is to clear the field to plant another crop. This is a common practice in market-gardens. Subsistence farmers may harvest part of their crop early because they need food before the main harvest. The undisturbed tubers continue to bulk until they are harvested later. If severe outbreaks of disease or insects occur, early harvest is a way to salvage the crop. Farmers who produce seed potatoes may harvest early to avoid virus buildup.

Some farmers choose to harvest late. If no other crop has to be sown, if the soil is free from pests, and if theft is not a problem, some farmers leave potatoes in the ground after harvest as "in-ground storage" for later consumption, sale, or use as seed.

Many highland farmers store part of their harvest in their homes or in general-purpose outbuildings for later consumption or use as seed. Few farmers invest in specialized storage structures for potatoes intended for market because potato harvests are normally staggered in mountainous areas, and it is difficult to predict future prices. For seed, however, many farmers, particularly in market-garden areas, are constructing specialized storage facilities.

Traditional potato processing systems, which take advantage of cool temperatures and high solar radiation, are found in the Andes and the Himalayas. These systems are based on the use of culls, surplus potatoes, or bitter varieties, which have little economic value unless they are processed. The processed products, however, lack wide consumer appeal, and use of these traditional processing systems seems to be declining. Large-scale industrial-type processing of potatoes is uncommon in highland areas.

Lowland potato production

Until recently, few farmers in the tropical lowlands grew potatoes. Varieties were not adapted, systems for supplying seed were rudimentary, and facilities for storing potatoes in hot areas seldom were available. In addition, cultural practices appropriate for potato production in warm environments were not widely known.

Since World War II, socioeconomic conditions and technological advances have come together to make the arid, irrigated lowlands the developing world's most dynamic and productive potato-producing zone. High yielding, early-maturing varieties have come into use. Seed production and distribution systems have improved, cold storage capacity has expanded, and farmers have developed ingenious methods of keeping potatoes for a few months without refrigeration. Through applied research by specialists and trial-and-error testing by farmers, new cultural practices and cropping systems have evolved to exploit the potato's genetic potential under the ideal growing conditions offered by many lowland irrigated areas. The result has been an extremely rapid expansion of potato production.

Producing areas. Important lowland areas producing potatoes are in the Indo-Gangetic plain (Bangladesh, India, Pakistan), Egypt, the Red River Delta in Vietnam, southern China, Cuba, and the Peruvian coast. In many of these areas, potato is still more expensive to produce than other staple food crops, and it is consumed as a vegetable. However, due to the rapid pace of technological change, potatoes are becoming cheaper, and this trend is increasing the importance of the potato in diets of people in these areas.

The lowlands are beginning to be an important source of potatoes for large urban markets, and in some countries, they generate a large share of the national potato production. In Peru, coastal producers supply half the potatoes consumed in Lima's metropolitan area. In India and Pakistan, over half the total production now originates in the plains. And in Bangladesh, which has no mountainous areas, potatoes have become the most profitable winter crop.

Cropping patterns. In the lowland tropics and subtropics, potatoes are typically produced on well-drained soils during the dry winter season with irrigation or in areas where residual moisture is abundant. Potatoes require a period of at least 60 days during which nighttime temperatures must remain below 20 degrees C. Environmental conditions are more uniform in such lowland areas than in the highlands. A wide range of crops are grown, depending on local market conditions, and farms vary greatly in size. Most potatoes are produced for the market. Due to the rather uniform environment and the market orientation of producers, potato technology is less varied in the lowland tropics than in the highlands. Potato yields are generally higher, and yields on small farms may equal or exceed yields on large farms, except where institutional arrangements limit small farmers' access to credit, markets, or new technologies.

Because potatoes have a short growing season, high yield, and high market price, they are an attractive winter crop in lowland tropical areas where soil moisture is adequate, pest problems are controllable, and nighttime temperatures are low enough to ensure tuberization.

Varieties. Until recently, no variety had been bred for the lowlands. Varieties that stored well grew poorly in hot climates, and vice versa. Fast-growing varieties are preferred by most farmers in areas where climate limits the growing season or where other profitable cropping options are readily available. Farmers who hold potatoes in unrefrigerated storage structures usually want varieties that have a long dormancy period. Unfortunately, many of these varieties also have a long growth cycle. In general, farmers are quick to adopt varieties adapted to hot growing conditions when they become available.

Seed systems. The shortage of good quality seed at a reasonable price has hampered potato production in lowland areas. Because of the high insect populations that prevail in lowland areas, locally produced seed becomes severely infected with virus diseases in the field or in storage. Furthermore, farmers have difficulty keeping stored seed potatoes in good condition in hot areas. As a result, many farmers choose to bring in seed from higher areas or from abroad.

Seed production in lowland areas is demanding technically and institutionally. It requires a well-organized program with trained personnel and facilities for aphid monitoring, laboratory tests, seed inspection, storage, and distribution. Local research also must be undertaken to understand the dynamics of insect populations and solve problems of virus degeneration and seed physiology.

Successful seed programs have contributed to the rapid expansion of potato production in several lowland areas. In the 1960s in Mexico, for example, the seed program linked lowland producers of consumer potatoes to seed growers in high areas. In India, an effective lowland seed system was developed that broke the dependence of plains farmers on seed transported from the hills. Making reasonably priced, healthy seed available at the optimal planting time stimulated expansion of potato-producing areas and led to better yields in lowland areas.

Lowland potato farmers normally plant at higher seeding rates than farmers in mountain areas. In irrigated lowland areas, crop yields are more responsive to input use than in rainfed, highland areas, and production risks are also lower. Where tuber-borne diseases (particularly bacterial wilt) occur, farmers generally plant whole seed tubers. If farmers want to break dormancy and hasten emergence of a more uniform crop, they often cut up the seed tubers.

Cultivation. Farmers in lowland tropical areas grow potatoes in the cool winter season, whereas farmers in subtropical areas may also grow potatoes in spring or fall. Large-scale farmers tend to use tractor-powered equipment for land preparation, but many small-scale farmers use draft animals. Planting, weeding, cultivation, and harvesting are usually done by hand, except if labor is scarce. Tractor- or animal-drawn lifters are sometimes used during harvest. Farmers usually plant potatoes in rows as a monoculture.

Farmers in lowland irrigated areas apply livestock manure, fertilizers, and pesticides heavily because of the high yield response and the severe pest and disease problems that usually exist.

Production costs. Good quality seed is difficult to find in lowland areas even if seed is being imported. For this reason, seed is generally the costliest input in lowland potato production (as shown previously in Table 15). As yields are highly responsive to fertilizer in irrigated lowland areas, fertilizers generally rank second as an input expense. The impact of pesticides on production costs is still small in most potato-growing lowland areas; however, in a few places they are becoming significant.

Harvesting and post-harvest technology. Farmers harvest early if another crop can be planted after potatoes, if they expect prices to fall, or if pests or adverse weather are likely to cause losses. Farmers rarely delay the harvest past senescence in lowland areas because in most areas another crop can be immediately planted and potatoes do not keep well in warm soil.

Because potatoes are usually a winter crop in lowland tropical areas, harvested tubers have to be stored and marketed during the hot months. Heavy losses may occur during handling, transporting, and storage. Until recently, farmers in hot areas could not hold potatoes for long periods, so prices plunged at harvest time, surpluses accumulated, and many potatoes spoiled before they could be disposed of. Producers often lost money on the crop, and consumers never established the habit of consuming potatoes because stocks disappeared from the market within a few months after harvest.

Storage difficulties also limited the ability of farmers to keep their own seed from one season to the next, making them dependent on seed from highland areas or from abroad. This made seed costly and seldom available in the optimal physiological stage when farmers wanted to plant.

Farmers have developed ingenious systems for short-term storage. Because the main storage problem in lowland areas is heat, most systems involve holding potatoes in dark, well-ventilated places in houses, under trees, or in specially designed outbuildings. If humidity is low, farmers may sprinkle water on the walls or floor to reduce storage temperatures. Building materials vary greatly. In Egypt, for example, permanent structures are made of mud bricks, whereas in coastal Peru, temporary storage facilities are made of reeds and thatch.

In traditional unrefrigerated storage, substantial losses may occur after a few months. Consequently, the introduction of refrigerated storage that allows potatoes to be held for several months has strongly affected potato production and use in many lowland areas. The advent of cold storage has been a key factor permitting production of seed potatoes in the Indian plains. Expansion of cold storage in Bangladesh has made potatoes available to consumers for a longer time after harvest. But cold storage is expensive to build as well as to operate and maintain. Occasional power failures and poor management often cause heavy losses. Most cold storage facilities have been built in or near major cities, and consequently, their benefits have accrued primarily to urban consumers and nearby producers. Distant areas, where poverty and malnutrition are usually most severe, have been little affected by this type of capital-intensive technology.

Potato processing has not developed to a significant extent in lowland areas.

Temperate-zone production

Producing areas. Temperate zones constitute the third major environment in which potatoes are grown in developing countries. The principal temperate potato-producing zones are in northern

Asia (China and North and South Korea) and South America (southern Chile and southern Argentina). Afghanistan, Iran, Pakistan, Syria, and Turkey also have extensive potato production in temperate areas. Potato growing takes place in summer and is dependent on rainfall, occasionally supplemented with irrigation. A few areas grow potatoes on residual soil moisture.

Because of the similarity of growing conditions, potato production systems in the temperate zones of some developing countries and in industrial nations are comparable. Potato farms in southern Argentina and Chile, for example, are similar to those in the northeastern United States and West Germany. However, differences also exist, particularly in Asia. This is due to the varied climates of temperate zones in developing countries as well as to differences in the socioeconomic context of potato production—patterns of economic and institutional development, input prices, the marketing system, and economic policies. For these reasons, many technologies employed in Europe and North America find little application in temperate zones of the developing countries like China.

Cropping patterns. The length of the growing period in temperate zones depends on latitude, elevation, and microenvironmental conditions. In some places potatoes are relay-cropped. In South Korea, for example, an early potato crop is grown before or after the main rice crop; in northern China, potatoes and maize are relay-cropped. In southern Chile and Argentina, potatoes are grown in rotation with cereals and pasture crops. In China, potatoes are grown both in extensive, rainfed field-crop systems and in irrigated vegetable gardens around towns and cities. While sole cropping is the norm in field cultivation, mixed cropping is common in vegetable gardens. Row cropping predominates in both systems.

Varieties. European or North American varieties are grown in many temperate zones. However, Japanese varieties are found in North and South Korea, and Russian varieties in China. In most temperate developing areas, potato breeding programs have now been established to select new varieties adapted to local conditions. In China, where many potatoes are used for manufacturing starch and noodles, breeders are looking for new varieties that have high dry-matter content. Two distinct types are sought: short-cycle varieties that permit relay cropping in vegetable systems and long-cycle varieties that produce higher yields in field cultivation systems. Plant breeders in South Korea place highest priority on identifying early-maturing potato varieties that can be relay cropped with rice. Drought- and virus-resistant varieties are desired in Turkey, where in warm, dry areas, potato yields are limited by moisture stress and virus diseases.

Seed systems. A fundamental difference exists between the seed systems of temperate zones and those of the tropics. In tropical areas, seed tubers move vertically, as a rule, from higher seed-producing regions to lower areas that produce consumer potatoes. In temperate areas, seed usually moves longitudinally, from north to south in the northern hemisphere and from south to north in the southern hemisphere. These seed movements in both environments have the same rationale: Potato growers seek to acquire planting material that has a low level of virus infection and is in good physiological condition. Hence, they look for seed produced in cool zones with low insect populations. The best seed in China comes from the northern provinces of Inner Mongolia and Heilungkiang. In Chile, the southern coastal areas near Puerto Montt produce the best seed.

Government seed certification programs patterned on those in Europe have been established in most developing countries that have temperate potato-production zones. The record of these programs has been mixed. Typically, after an initial period in which basic stocks of clean seed were produced, diseases have appeared in later multiplication generations, and as a result, farmer demand for certified seed has been weak.

Cultivation. In temperate zones, fields are cultivated with draft animals or tractors, depending upon the level of technological development and the relative costs. Tractors, for example, predominate in southern Argentina and draft animals in northern China. Hand cultivation is used in market-gardening areas everywhere.

As in the tropics and subtropics, most farmers apply livestock manure to potato crops. Chemical fertilizer is also widely used. Late blight is the most common disease in temperate areas, and application of fungicides is widespread. Insect problems are less severe and insecticides are less heavily used than in the lowland tropics. There are exceptions, however. In warm, dry areas, such as eastern Turkey, late blight is not a problem, but insects are.

Production costs. Temperate areas generally produce their own seed potatoes, so, although seed remains a major expense, it is less significant than in areas that import seed (see Table 15). The relative importance of implements and labor and other forms of power (draft animals, tractors) depends on local conditions. For example, potato production in southern Chile is mechanized to a much greater extent than in North and South Korea. Fertilizers usually account for a substantial share of production costs, but pesticides do not.

Harvesting and post-harvest technology. In highly developed areas, like Valcarce in Argentina, harvesting and grading is mechanized. Elsewhere, implements drawn by horses or oxen are used to unearth potatoes, which are then sorted and bagged by hand. Horses or oxen are used to draw wagons in less developed areas, and trucks or tractors are used in the more advanced areas.

Potato storage presents fewer problems in temperate areas than in most tropical and subtropical zones because harvested tubers usually are healthier and the storage period is cool. The main problem is to keep potatoes from freezing rather than from sprouting, drying out, or rotting, as is the case in the tropics. For this reason, potatoes are often kept in root cellars, farmers' houses, livestock stables, or other warm places.

In some temperate areas far from urban population centers, potatoes are cheap enough to be used as a raw material in starch production. In southern Chile, for example, a few large firms buy potatoes in bulk and sort them into three categories. The largest tubers are shipped to the consumer potato market in Santiago; intermediate-sized tubers are shipped to the Central Valley to be used as planting material; the smallest tubers and the culls are processed into starch. In northern China, potatoes are also used for starch production. This results from the high cost of transporting fresh potatoes south to Beijing and other large cities and also from governmental restrictions on commodity trade across provincial boundaries. These restrictions have recently been relaxed, and it is expected that in the future, fewer potatoes will be used for starch manufacture in China.

Mediterranean potato production

Potato-producing areas with Mediterranean climate (dry summers and cool, humid winters) are concentrated in North Africa and the Middle East. Outside of this region, the only developing country with extensive potato production under Mediterranean conditions is Chile. Because potato production systems in Mediterranean areas are in many respects intermediate between those of temperate and lowland tropical zones, only the significant differences are outlined here.

The Mediterranean climate often permits both a spring and fall potato crop. In areas like Cyprus, Tunisia, Algeria, and Morocco, seed for the spring crop is imported and a part of the harvest is used for planting the fall crop. Seed tubers often arrive from Europe in the apical dominance stage, and unless desprouted, they produce low-yielding crops. Storage of seed from the spring harvest until the fall planting is difficult because of high temperatures that occur during summer.

Farmers in Mediterranean zones generally plant European potato varieties, and some European breeders now test potential new varieties in North African countries before releasing them.

In parts of North Africa and the Middle East, the potato tuber moth has become a major pest, both in the field and in storage. Many farmers apply heavy doses of insecticides to control the tuber moth, but insecticide use in storage can be dangerous for consumers. Research and extension efforts are focusing on use of safer insecticides and alternative insect control measures, like more careful hilling (to keep the insects from reaching the tubers) and the use of pheromones in monitoring insect populations to alert farmers when it is necessary to spray.

Examples of production systems

This section presents six examples of potato production systems. The Mantaro Valley, Peru, illustrates field crop potato production in the high Andes. Cañete Valley on Peru's coast and Bogra and Tongibari in Bangladesh illustrate lowland tropical and subtropical potato production in the winter season.

Benguet Province, the "salad bowl" of the Philippines, illustrates market-gardening in a mid-elevation Asian zone. Northern Rwanda illustrates subsistence potato production in the central African highlands. Southern Chile illustrates potato production in a temperate zone.

The Mantaro Valley, Peru

In recent years, the population of Peru has grown faster than the output of potatoes, so per capita consumption of this staple food has fallen to about 65 kg a year. In the highlands, consumption nevertheless remains high, averaging about 130 kilograms per capita.

The Mantaro Valley, which is located in the central highlands about 200 kilometers from Lima, the nation's capital, provides an example of the complex potato production systems that characterize the Andes. Good road and rail connections make the valley a major supplier of potatoes to Lima and other coastal markets. Farmers have access to a wide range of agricultural inputs in Huancayo and several of the valley's smaller towns and villages. A university and a government experiment station in the valley have conducted potato research and operated extension programs there for several decades. Also in the valley are the highland experiment station of the International Potato Center (CIP) and the headquarters of the national potato program.

The valley's 150,000 hectares of cropland are in three distinct production zones: a low zone that is relatively flat along the Mantaro River from 3,200 to 3,450 meters above sea level; an intermediate zone, sloping land from 3,450 to 3,950 meters; and a high zone, yet more steeply sloping fields between 3,950 and 4,200 meters (Fig. 14). Land above 4,200 meters is used for pasturing sheep, llamas, and alpacas.

The production practices and technology needs of farmers on the valley floor differ markedly from those of farmers cultivating potatoes on the slopes. Moreover, within each zone, the practices and technology needs of the small subsistence-oriented producers

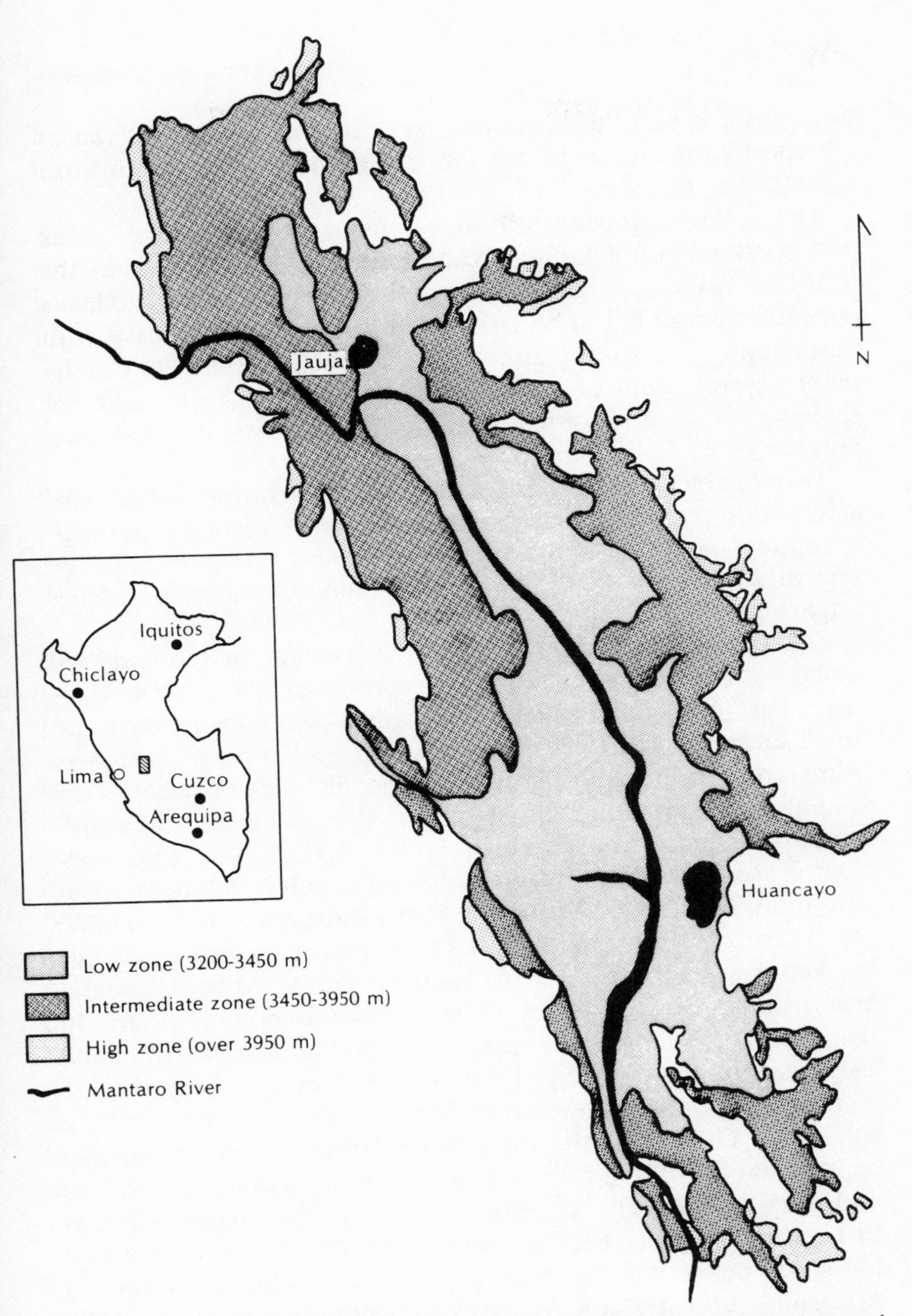

Figure 14. Agroecological zones of the Mantaro Valley, Peru. *Source:* Adapted from E. Mayer, *Land use in the Andes* (Lima: International Potato Center, 1979).

differ from those of the large commercial farmers. The differences are most pronounced in the low zone where extremes in farm size are the greatest.

The valley's growing season begins with the October rains and continues until killing frosts occur in May. Land near the Mantaro River is irrigated, but most cultivation in higher fields depends on rainfall. The risks of frost and hail increase with altitude, but pest and disease problems decline. Farmers usually plant several plots of potatoes (the average is 4) in different ecological niches in order to minimize the risk of total crop failure.

Two distinct potato production and marketing subsystems exist within the valley: one is small-scale and subsistence-oriented. It supplies potatoes to the valley's rural and urban population. The other is large-scale and commercially-oriented. It supplies Lima and other distant markets.

The typical small farm has only 1 hectare of cropland, of which 0.2 hectares is in potatoes. In contrast, the average large farm, has 75 hectares of cropland with 42 hectares in potatoes. Small farms are scattered throughout all ecological niches; large farms are concentrated in the low and high zones, where they specialize in commercial potato production and livestock raising. Small-scale farmers operate highly diversified, risk-averting, part-time farming systems and grow potatoes mainly for home consumption. Most small farmers derive a substantial part of their total earnings from off-farm jobs in mines, construction, or marketing, or on larger farms in the vicinity. Although nearly every farmer in the valley cultivates some potatoes, just 10 percent of the farms produce over half the potato crop, and they generate an even higher share of the market supply.

Throughout the valley, farmers grow potatoes in rows as a sole crop. They plant the rows about 1 meter apart to provide enough soil for hilling the varieties, which have abundant foliage and long stolons. The seeding rate varies from about 0.5 tons to nearly 2 tons per hectare, depending on seed tuber size and the space between hills in the row. Seeding rates are lowest in high areas and heaviest in irrigated fields on the valley floor. Border rows of "barrier" crops like lupines (*Lupinus mutabilis*) are sometimes planted to reduce damage from pests and livestock.

TABLE 16
Potato Production Characteristics in the Mantaro Valley, Peru, by Agroecological Zone, 1977/78

	Proportion (%)				
Zone	Potato producers	Potato area	Potato production	Potato area vs. total cropland	Average yield (t/ha)
Low	51	49	55	19	5.5
Intermediate					
East	24	30	31	39	5.0
West	18	13	7	22	2.7
High	7	8	6	57	3.6
Whole valley[a]	100	100	100	25	4.8

[a] Total may not add to 100 due to rounding

Source: E. Franco, D. Horton, and F. Tardieu, *Producción y Utilización de la Papa en el Valle del Mantaro, Perú* (Lima: International Potato Center, 1979).

In the low zones, crop production is intensive and maize is the predominant crop. Progressively higher on the slopes, maize and several other crops disappear, and fallow plays an expanding role in the rotation. Andean tuber crops (mainly potatoes, but also *mashua* and *oca*) predominate on the humid eastern slopes of the valley; small grains (mainly barley) predominate on the drier western slopes. Potato is the main crop grown in the high zone (Table 16).

Native potato varieties predominated until about 1950, when farmers in the low zone began to adopt varieties released by Peru's Ministry of Agriculture. In the higher areas, however, modern varieties still do not have a substantial yield advantage over native varieties for most farmers. By growing native varieties that are resistant to frost and hail without chemical fertilizers and pesticides, small farmers in high areas achieve a stable food supply while limiting their production costs and risks. Native varieties are also preferred by consumers; they store better than modern varieties, and they sell at a premium price.

About three quarters of the farmers in the intermediate and high zones plant seed tubers that they have stored from the previous season. In the low zone, less than half the farmers use their own seed. Most obtain seed from higher areas because it yields better than seed tubers grown and stored in the low zone.

The government has attempted to set up a seed certification system in the valley and has produced a small amount of "guaranteed" seed. Although a few of the large growers have purchased this seed, other farmers continue to use the common seed available in the region. On-farm research has shown that the yield advantage of the guaranteed seed is insufficient to be profitable for most small farmers in the valley.

Land preparation and tillage practices vary among locations (zone, relief, soils) within the valley and among the different types of farms. In the low zone, most large farms cultivate with tractor-powered equipment, whereas small farms use oxen. In higher areas, tractors are occasionally used on relatively flat parcels of land near roads, oxen are used on moderately sloping land, and hand tools are used in steep fields. All farmers fertilize potatoes with livestock manure, but only fields in the low zone receive near the recommended levels of chemical fertilizers and pesticides. In high areas, these inputs have a smaller and more variable effect on potato yields. As their use increases financial risks, application rates are far below recommended levels.

Potatoes are usually planted in October and harvested in May. Along the river, farmers plant and harvest earlier on irrigated, stony fields that are less susceptible to night frosts. Farmers at higher elevations also plant a few protected fields early in order to harvest potatoes when food is in short supply and prices are at their peak. The threat of heavy losses from soil-borne pests precludes delayed harvesting for most farmers.

Farmers store some potatoes for household needs and for seed, but storage for later sale is not common because harvests elsewhere in the country can depress market prices. Most farmers hold potatoes in their houses, which allows them to keep a close eye on the stocks and minimizes their capital investment in storage facilities. Some prosperous farmers store potatoes in multiple-purpose outbuildings. A handful of large farmers keep their seed potatoes in specialized seed potato stores.

Food processing has historically been an important economic activity of rural Andean households. Small farmers process potatoes into several dehydrated products, such as *chuño*—a freeze-dried product made from bitter varieties (*Solanum juzepczukii* and *S. curtilobum*)—and *papa seca*—a cooked and sun-dried product made from common varieties. Market demand for chuño and papa seca is limited. Chuño has little acceptance outside the high Andes, and most Peruvians consider it to be a low-prestige food.

Local researchers and extension agents need to understand the valley's different production systems in order to address the technology needs of diverse groups of farmers. Farmers in high areas, for example, want frost-resistant potato varieties with higher and more stable yields and low input requirements. Farmers in lower areas are less concerned with frost hazards than with insects and nematodes, seed quality, and potato storage. They are also more willing to use inputs to maximize yields. In the low zone, where the yield gap between large and small farms is substantial, a more aggressive and effective extension program is needed to raise the productivity and incomes of small farms. In its absence, the trend toward concentration of landholdings in a few large farms will continue.

The Cañete Valley, Peru

Located in the central coastal region, 150 kilometers south of Lima, the Cañete Valley provides a counterpoint to the potato production systems of the Mantaro Valley. The entire valley is irrigated, and its climate and soils are ideal for a broad range of warm-weather crops. Cañete's transportation and communication facilities are excellent. Most of the land was in large sugarcane plantations until world sugar prices collapsed during the great depression. Then cotton emerged as the valley's dominant crop. Potatoes have been grown in Cañete for many years, but rapid expansion of potato production began only after World War II.

Changes in demand and supply have altered Peru's age-old production systems. Lima's rapid growth has created a large-year-round market for potatoes, but the central highlands can adequately supply Lima with potatoes only from March until

July. The escalation of potato prices in other months has stimulated winter production on the coast.

In addition, since around 1950, several new short-cycle potato varieties have been released that are highly responsive to fertilizer and yield well on the coast in winter. An informal seed system has emerged to supply coastal growers with quality highland seed of these varieties. Many production inputs, including chemical fertilizers, pesticides, and machinery, have also become readily available, and the Ministry of Agriculture has expanded its extension service in the region. As a result, Peru's coastal potato production grew from 25,000 tons in 1950 to 150,000 tons in 1980. Although the coastal valleys account for only about 10 percent of the national potato production, they supply nearly half of the potatoes consumed in Lima.

The valley has a pronounced summer season, from January to March, with hot sunny days and temperatures ranging from 18 to 27 degrees C. The cooler, cloudy winter season, from May to November, has temperatures ranging from 15 to 21 degrees C. Cotton and maize are the principal summer crops; potatoes are the principal winter crop, along with sweet potatoes, forages, and pulses.

Two potato crops are grown: the early crop is planted around April 1 and harvested around August 1. It accounts for about a fifth of the valley's potato production. The main crop is planted around June 1 and harvested around November 1. Yields in the early season are normally 15 to 20 percent below those in the main crop. However, this difference is often more than offset by the higher prices farmers obtain for early potatoes.

The Peruvian potato breeding program has been successful in introducing new varieties on the coast, in part because varieties quickly become infected with virus diseases and their yield potential drops. Cañete's potato crop is grown from varieties released since the war; native Andean potato varieties are entirely absent.

Farmers plant a part of the early crop with seed stored from their previous harvest, but the main crop is planted with seed tubers trucked from the central highlands. Farmers prefer highland seed because it produces more vigorous and higher-yielding crops. Due to the seasonal planting and harvesting cycles of the two zones, however, highland seed in proper physiological condition is not available when coastal farmers have to plant the early crop. In the 1960s, in an attempt to overcome this problem, a group of the valley's large farmers constructed a refrigerated facility for storing seed potatoes through the summer. With the land reform of 1968, this facility fell into disrepair and disuse. Some farmers now store small tubers from their main-season harvest in shaded, naturally ventilated structures until the following year's early planting. But the stored tubers are vulnerable to desiccation and insect damage, and they often are carriers of virus diseases.

Most farmers plow with tractors. On large farms, tractors are also used for hilling and spraying pesticides. Small farmers tend to use draft animals for hilling, and they apply pesticides with backpack sprayers. All farmers apply organic material and chemical fertilizers on potatoes. The average application is 10 tons of decomposed livestock manure, 300 kilograms of nitrogen, 150 kilograms of phosphate, and 150 kilograms of potash per hectare. These application rates exceed official recommendations.

Late blight and insects, primarily leaf miner fly, do considerable damage to the crop, particularly when winter weather is mild. Consequently, potato growers frequently apply fungicides and insecticides, usually together.

Cañete farmers overcame severe insect problems in cotton in the early 1960s through integrated biological control measures, including rigid planting and harvest dates, burning of trash, monitoring of pest populations, and reduced spraying. Instituting a similar program with potatoes today could be hampered by the larger number of producers and the fragmentation of land holdings.

Harvesting generally begins in July and continues until December. The entire early crop and most of the late crop are marketed immediately after harvest. Less than 10 percent of the harvest is retained for seed, and practically no potatoes are kept for home consumption. Farmers prefer to buy their potatoes fresh in the market.

The two most pressing technology needs of Cañete's farmers appear to be a reliable supply of high-quality, reasonably priced seed, particularly for the early season, and cost-effective and ecologically sound methods of controlling insects that attack the potato crop.

Tongibari and Bogra Sadar, Bangladesh

Potato production has expanded rapidly in Bangladesh in recent years, making the potato the second most important food crop in the winter, or *boro,* season.

Bangladesh is one of the most densely populated countries of the world. The population, 95 million, is principally agricultural, with over 80 percent living in rural areas. Cultivated land totals only about 9 million hectares, giving a rural population density of 9 persons per hectare.

Cropping patterns are dominated by the monsoon, which floods much of the country from June to October. In earlier times, nearly all crops were produced during the monsoon season, with rice and jute the leading crops. Over time, tube wells and low-lift pumps have been installed, land improvements made, and new crops including modern varieties of rice introduced to expand production in the winter season. Potatoes, wheat, and "boro" rice have become the principal winter crops. Other vegetables, pulses, spices, and tobacco are also grown during the winter on Bangladesh's rich deltaic soils.

Potato production in the late 1940s was about 200,000 tons on 35,000 hectares. Since then, production has increased to over 1 million tons, the area under potatoes has nearly tripled, and yields have increased by about 50 percent.

Expansion of potato production in Bangladesh has been intimately associated with four factors: introduction of modern varieties, mainly from the Netherlands and India; importation of clean seed stocks; development of local seed production capacity; and expansion of the cold storage industry.

At present 50 to 60 percent of the total area under potatoes is covered with what are known as "modern varieties" (mainly Dutch) and with "local varieties" from earlier introductions. Many of these are of the subspecies *andigena* and produce a large number of small tubers. Because local varieties are contaminated with viruses, they yield poorly. Their culinary and

Traditional potato varieties in Bangladesh are rather low-yielding and produce large numbers of small tubers, but they give rather stable yields and store well.

keeping qualities are good, however. These varieties are grown mainly in areas where farmers do not have access to cold storage because their tubers can be stored without refrigeration (in baskets, for example) from harvest until the next planting season.

Potato production using modern varieties has developed mainly near large consumer markets where cold storage facilities have been erected. As seed stocks degenerate rapidly due to viruses in Bangladesh, they need to be replaced frequently. In the 1970s production of modern varieties required 4,000 tons of imported seed each year. The Bangladesh Agricultural Development Corporation (BADC) has developed a program for local seed production with technical assistance from the Netherlands. The aims are to increase local seed production and to improve quality, thus making it possible to reduce imports. Seed is being produced on a contract basis by individual farmers.

The program has made good progress in producing more and better seed. In 1980 about 2,600 tons of seed were imported and the BADC program produced only 1,700 tons. By 1985 seed imports were reduced to 1,200 tons and the BADC was producing 4,000 tons. The quality of locally produced seed has also been improved. In 1980 barely 60 percent of the BADC seed lots had less than 2 percent of virus, but by 1985 the proportion had increased to nearly 80 percent.

The BADC program, which is now operating in 10 zones, plans to expand annual seed production to between 5,000 and 7,000 tons over the next few years.

Before the advent of cold storage, farmers had difficulty keeping seed from one season to the next, and potatoes were on the market for only a few months after harvest each year. Refrigerated warehouses began to be constructed in the 1960s, and during the 1970s, cold storage capacity tripled to 280,000 tons. By the mid-1980s there is excess capacity around Dhaka and other urban centers, but in potato-producing areas, where a larger proportion of the warehouse space is devoted to seed potatoes, insufficient cold storage capacity still limits potato production.

Bangladesh is a small country with no mountains and an average elevation of only 10 meters above sea level. Yet, cropping patterns and technology vary between regions, depending on local variations in soils, intensity and duration of the monsoon, relief, and economic conditions. This is illustrated by a comparison of potato production systems in two areas: Tongibari, in central Bangladesh; and Bogra Sadar, about 100 kilometers to the northwest.

Farms in Tongibari have on average 1.6 hectares of cultivated land in 11 distinct plots. Most plots are triple cropped, the dominant crop sequence being rice, potato, and sesame. The principal crops in the locality are rice in the monsoon season and potatoes in the winter. The average farmer cultivates 0.8 hectares of potatoes in six plots measuring only 0.15 hectares each.

Two-thirds of Tongibari's farmers keep milk cows, but only about 10 percent have draft animals. Four-fifths of the farmers prepare their land with power tillers, in some cases supplemented with rented bullocks. The power tiller is not common in Bangladesh, but the high profitability of potato growing and the difficulty of maintaining draft animals (due to annual flooding) have made the power tiller popular in this area. Plowing and planting is usually done in November and December.

The Dutch varieties, Multa and Patrones, are grown by the vast majority of farmers in the area. Three-fourths of the farmers plant seed from their previous crop, which they have held in rented cold storage space. A few farmers purchase imported seed, which they multiply for their own use and to sell to other growers or consumers.

Tongibari farmers apply an average of 7 tons per hectare of livestock manure on their potatoes, and nearly all use chemical fertilizers and pesticides. Average fertilizer applications are above the recommended levels. Pesticides are used mainly to control late blight, which is common in the area but seldom severe. Only about 10 percent of the farmers irrigate their potato fields. To reduce soil temperatures, control weeds, and preserve moisture, many farmers spread rice straw on the fields. Weeding is practiced on only about half the farms.

Potatoes are manually harvested during February and March, about 90 days after planting. Yields are reported to be over 20 tons per hectare—double the national average. Yields are normally higher on larger farms that use high seeding rates and more fertilizer and have their crop in the ground for a longer time. About two-thirds of the harvest is placed in cold storage for sale later in the year; 20 percent is sold at harvest, and the balance is kept for household consumption or held in cold storage for use as seed the next year.

In Bogra Sadar, the potato production system differs in several important respects from that in Tongibari. Farms are smaller, averaging only 1.1 hectares of cropland and 0.2 hectares of potatoes in two plots. Double cropping prevails, the dominant crop sequence being jute, fallow, and potato. Eighty percent of the farms own draft animals and the rest hire them for plowing; power tillers are not used. Most farmers grow a single potato crop each winter, but one out of five farmers plants two consecutive potato crops in the same plot during the winter season. The first planting begins in September for harvest in late November and December, when potato prices are at their highest. A second crop is planted in December and January and harvested in March. The average duration of the first and second crops is 60 and 80 days, respectively. Farmers produce the first crop mainly for cash. Part of the second is used for household consumption and seed. Two consecutive crops are possible because the land is higher and the monsoon shorter than in Tongibari to the south.

The main potato varieties in Bogra are the Dutch variety, Cardinal, and the Indian variety, Kufri Sinduri. Other Dutch varieties are also grown, along with several older varieties. Most farmers purchase seed rather than use their own because cold storage capacity in the region is limited. About one-quarter buy seed from the Bangladesh Agricultural Development Corporation. An important reason why some farmers in Bogra continue to grow older varieties is that they store better without refrigeration than the higher-yielding Dutch varieties.

Bogra farmers apply nearly 16 tons per hectare of livestock manure to their potato crops, but only about a third as much chemical fertilizer as the farmers in Tongibari. In addition, they all irrigate and weed their fields. Late blight and other pest and disease problems are less serious in Bogra, and only about 60 percent of farmers apply pesticides.

Due to the longer growing season and staggered planting dates in Bogra, potato harvesting is spread from November to March. Yields average about 13 tons per hectare, giving a total potato production of about 3.5 tons per farm. Ninety percent of the production is sold immediately after harvest. The remainder (only 350 kilograms per farm) is kept for household consumption and seed or is sold later in the year.

In both areas, seed is the largest production expense, accounting for 40 to 45 percent of the total production cost. Chemical fertilizer is the second greatest expense in Tongibari, labor in Bogra. The potato crop requires an average of 370 days of labor per hectare in Bogra, 270 in Tongibari. Pesticides account for less than 4 percent of the production cost in both areas. Total production costs per hectare are one-third higher in Tongibari than in Bogra, but due to larger yields, production costs per ton in Tongibari are 40 percent lower. Bogra farmers receive higher prices for their output (due to sales in November and December when prices peak), but still their net returns per hectare, and particularly per day of household labor employed, are significantly lower than those in Tongibari.

Farmers in these two regions in Bangladesh have some technological needs in common whereas others are site-specific. In both areas, seed costs are high, and good quality seed is scarce. Increasing the supply of high-quality seed would benefit producers and consumers in both areas. Expansion of cold storage capacity in Bogra could help improve the seed supply and bring down production costs. In both areas, yields might also be increased and unit production costs reduced through introduction of new varieties that are better adapted to local conditions. For example, quick-bulking short-cycle varieties, would be useful for the early planting in Bogra when two crops are grown. In contrast, Tongibari farmers can make best use of somewhat later-maturing varieties. Consumer preferences also differ between the regions. In central Bangladesh consumers prefer white-skinned potatoes, in the north they prefer red-skinned ones.

Given Bangladesh's unique, lowland tropical environment, it seems likely that present potato yields are far below the economic ceiling, and that applied research and production programs in the country's major production zones could generate high social returns through rapid increases in yields and reductions in potato prices. Increased availability of high-quality seed of desired varieties and improvements in irrigation would probably have the largest impact on production.

Benguet Province, the Philippines

Benguet Province, in northern Luzon, is typical of highland vegetable-producing areas in Southeast Asia. Vegetables have been the principal farming enterprise in Benguet Province since the area was cleared from natural forests earlier in this century. Together with a small area in neighboring Mountain Province, Benguet accounts for over four-fifths of the Filipino potato production. National production is currently estimated to be about 40,000 tons, giving a per capita consumption of less than one kilogram. Approximately 12,000 farmers in the area grow potatoes. Cultural practices have changed rapidly since World War II.

A mountain range bisects Benguet Province from northeast to southwest, with numerous agricultural valleys along it. Most of the terrain is broken with steep slopes and ravines. Most soils are deep and well drained. There are two distinct seasons: a wet monsoon season from late April until November and a dry season from December until April. Annual rainfall averages 4,000 millimeters of which 60 percent occurs between July and September. The maximum daily temperature averages 23 degrees C; the minimum averages 15 degrees C. Heavy rains, winds, and typhoons occur from July to September, and hail is a hazard during May and early June. Night frosts may occur during the dry months, particularly in January.

Commercial production of cabbage, carrots, potatoes, and tomatoes began in the area in the 1920s when extension of a road north of Manila opened forest lands and stimulated migration to the north. Following World War II, land clearing and settlement continued northward along the mountains and outward into the adjacent valleys. Today vegetable production extends for some 120 kilometers north of the city of Baguio. Potatoes have gradually replaced cabbage as the main cash crop. Recently, potato cultivation has also extended into rice paddies following the rice harvest in December.

Vegetable farms range from about 0.3 to 4.0 hectares, with the average just over 1 hectare. Most farms are family owned and operated. Many farmers, especially the larger ones, must hire labor because all farming operations are done by hand. Many vegetable farmers maintain strong links with trading families or engage in trading themselves.

Potatoes and cabbage, the main crops, are usually grown alternately, but some farmers grow the same crop for two or three successive seasons. Minor crops, such as carrots, snow peas, and Chinese cabbage, are grown in some areas. The considerable variation in planting and harvest dates between upland and valley areas ensures a steady supply of potatoes to markets throughout the year. The main cropping season is from March to July. Irrigation in some areas allows a second cropping season from October to February. Growing potatoes during the typhoon months, August and September, is risky, but a few farmers plant at this time to benefit from high December prices. In the valley-bottom areas where irrigation is plentiful, potatoes are planted in the dry season and harvested from December to March, when prices are at their highest.

What Benguet farmers call native potatoes have been replaced since 1950 by varieties brought in from abroad. In the 1960s, two Mexican varieties—Greta and Conchita—were widely adopted because of their late-blight resistance and high-yield capacity. They still account for over half of the wet-season plantings. But there have been no further importations of these Mexican varieties since their introduction, and they are now heavily infected with virus diseases. A German variety, Cosima, accounts for nearly 20 percent of the wet-season crop. Because it is susceptible to late blight, farmers spray it heavily with fungicides. An important reason farmers grow Cosima is the availability of high-quality imported seed. Small quantities of Red Pontiac are also regularly imported from the United States. A number of Dutch and German varieties have been tested on the experiment station and made available to farmers. These have not become popular, however, because they are susceptible to virus diseases and because there is no reliable supply of clean seed.

The farmer's choice of variety often depends more on the availability of seed than on varietal characteristics. For the wet-season crop, most farmers plant their own seed tubers, saved from the previous crop. Every 3 to 5 years, they purchase fresh seed from farmers at higher elevations to "regenerate" their stocks. Some farmers who have land at different elevations move seed each season from high fields to low fields, and vice versa.

The scarcity, high cost, and variable quality of seed is a major problem for potato producers. Researchers pressed for establishment of an official seed program as early as the 1960s. In

1977 a seed certification program was initiated under an agreement between the Philippines and the Federal Republic of Germany. Small quantities of high-quality seed are now being made available to growers.

Land preparation, like all other cultural practices, is done by hand. Farmers customarily grow vegetables, including potatoes, in double rows along beds measuring 70- to 90-centimeters wide, with a canal of some 40 centimeters between beds. Farmers plant each seed piece in individual holes spaced about 35 centimeters apart. At planting time, chicken dung, purchased from large poultry producers in the lowlands, is mixed with the soil. Most farmers apply over 5 tons of chicken dung per hectare. Chemical fertilizers are commonly applied at high rates as a side-dressing when potatoes are hilled up.

Where irrigation water is abundant, many farmers practice furrow irrigation. In the marginal areas, water is often applied once or twice a day using watering cans. Some farmers have jointly drilled wells or tapped springs; they install small metal or plastic pipes to bring water to each holding.

Late blight is the major disease of potatoes in the area; this accounts in large part for the continuing popularity of resistant Mexican varieties. Farmers sometimes spray European varieties as often as every third day during the wet season. When monsoon rains prevent spraying, the disease builds up rapidly and kills susceptible varieties.

Many farmers harvest immature potatoes to take advantage of high prices or to clear the field for another crop. They occasionally store part of their wet-season crop for a few months in the hope of receiving higher prices in October or November. The dry-season crop is seldom stored because prices are usually near their peak at harvest time. Farmers have other uses for their land at this time, and many soils are too warm and dry to delay harvest. The harvest of the wet-season crop, however, may be postponed by farmers who have well-drained soils. They may earth up the beds two or more times to cover tubers that have become exposed by rain. Losses during the first 3 months are small.

Although potato production systems vary, improving seed quality, seed storage, and blight control are three common concerns of most Benguet potato farmers. Economically attractive new practices have been rapidly incorporated into the area's already highly productive market-gardening systems.

The volcanic region of northern Rwanda

Situated in the highlands of central Africa, Rwanda offers an example of rapidly expanding potato production in a subsistence-oriented food system. More than 90 percent of Rwanda's population is engaged in agricultural production, distribution, and related services. Most agricultural production is subsistence-oriented. Only a small proportion of food crop production is marketed, and industrial crops account for only about 7 percent of the total value of agricultural production.

Potato production has nearly quadrupled over the last two decades, far outpacing the growth of total food production. Current potato production is estimated to be 240,000 tons on 37,000 hectares, giving an average yield of 6.5 tons per hectare and an estimated per capita potato consumption of 33 kilograms.

Potatoes are grown throughout Rwanda but are an especially important food crop at elevations over 1,500 meters above sea level. Most potatoes are grown along the Zaire-Nile ridge and on the northern volcanic slopes, where annual rainfall averages 1,200 to 1,400 millimeters and temperatures average 16 degrees C with considerable variation from day to night. These areas have a bimodal rainfall distribution, which permits two crops each year.

Most of Rwanda's cropland is in small family farms. Farms average 0.8 hectare in the high potato-growing areas of the northwest and 2.1 hectares in lower regions to the southeast. Farms are extremely fragmented: A holding of less than 1 hectare may consist of eight to ten small plots several kilometers apart. But by cultivating land in different ecological zones, farm families diversify their food supply and spread production risks.

Farmers attempt to produce enough food to satisfy their household requirements. They generally interplant potatoes with maize, beans, or cereals. Intercropping helps farmers intensify land use and produce a more continuous supply of food throughout the year. Following the bimodal rainfall pattern, farmers plant two potato crops each year; the main (or rainy) season crop is planted around October, the short (or dry) season crop around May. Most potatoes are harvested from February to April and in September and October. Fresh food in general is scarce during 3 or 4 months before the short season and 2 or 3 months before the main season. Food shortages may be acute from May to July. Fast-maturing crops such as potatoes, beans, and peas are valued because they provide a source of fresh food after periods of scarcity.

Land for the potato crop is prepared by hand. In many areas, farmers plant their crops on large ridges or mounds about a meter wide, 2 or more meters long, and 25 to 50 centimeters high. Farmers use this system to ensure adequate drainage during rainy periods and also to accumulate fertile soil. As a rule, potatoes are sown with an irregular spacing of about 20 to 40 centimeters between the plants. After planting, potatoes are hilled and weeded once or twice.

A number of varieties are grown. During the late 1960s blight-resistant Mexican varieties were introduced from Uganda. Sangema, which was introduced from Mexico in 1972 and multiplied by the Programme National de l'Amélioration de la Pomme de Terre (PNAP), has become a principal variety in northern Rwanda.

Farmers try to retain enough tubers from their own potato crop to plant the following season, but occasionally they must purchase seed potatoes on the local market. Before the recent establishment of a seed potato production and distribution system, farmers had to get seed potatoes from consumer markets.

Chemical fertilizer and pesticides are not readily available in Rwanda, but a few commercial farmers along the main roads have begun to use fungicides to control late blight. Other farmers use no chemical inputs, and to avoid outbreaks of late blight—the most important disease—they commonly delay planting to avoid heavy rainfall. The delay, however, results in lower yields because the growing season is pushed into a period when rainfall is less adequate. Bacterial wilt is present in all producing areas, but it is particularly costly for small farmers who like to leave volunteer plants as seed for the next crop and cannot afford to use long rotations that help reduce soil infestations of wilt.

Most potatoes are harvested about 4 months after planting, but premature harvesting is common to satisfy urgent food needs and avoid theft. Compared with potatoes grown in the short (dry) season, the quality of potatoes produced during the main (rainy) season is generally poorer due to irregular rainfall and infection with late blight. Main-season potatoes are also more susceptible to rotting in storage. Consequently, farmers store a smaller proportion of their main-season harvest. This practice makes food scarce in the following months and limits the amount of seed tubers available for the May planting. Farmers often store potatoes harvested in the dry season in their houses (to prevent pilferage) or keep them in large baskets in the housing compound.

In northern Rwanda, potato farmers are eager to obtain new varieties that are resistant to late blight and bacterial wilt. Until the input market situation changes drastically, chemical fertilizers and machinery cannot be expected to affect Rwanda's potato production significantly.

Southern Chile

Chile is an advanced developing country, with an average per capita income of US$2,000 and 80 percent of its population in urban areas.

Chile's 5 million hectares of cropland are scattered from the irrigated valleys in the arid, subtropical north to the cool, rainy temperate southern regions. Potatoes are grown in nearly all regions. After wheat, potato is one of Chile's most important crops. Forty percent of Chile's farmers grow potatoes, but the average farm has less than 1 hectare of potatoes and harvests only about 7 tons a year. Few farmers specialize in potato production because unstable yields and prices make returns from the potato crop highly unpredictable. Most farmers cultivate small plots of potatoes in rotation with other food and pasture crops in order to supply food for home consumption, feed for livestock, and supplemental cash income.

Potatoes are particularly important in the farming systems of southern Chile, which has a cold winter and a cool, rainy summer. The region is primarily a livestock-raising zone, with only a small part of its farmland dedicated to crop production. Where crops are grown, potatoes are the first crop in the rotation, followed by wheat, other cereals, or forages. About 60 percent of the farmers plow with tractors. The remainder, smaller-scale farmers or those with uneven land along the coast, plow with draft animals.

Until the 1950s when late blight first caused widespread crop failure, most potato growers planted a mixture of local varieties known as Corahilla. The Instituto Nacional de Investigaciones Agropecuarias (INIA), the national agricultural research institute, subsequently introduced more resistant European varieties to replace the susceptible varieties. At present, most farmers cultivate the varieties Desiree, Pimpernel, and Ultimus.

Potato farmers in southern Chile commonly plant about 2 tons per hectare of seed tubers that have been stored from the previous harvest. To minimize the buildup of virus infection in their seed stocks, farmers purchase small amounts of clean seed from recognized seed growers and multiply it for subsequent crops. The quality of most seed is far better than that of other developing areas. Virus infection is minimal in seed lots harvested along the sea, where aphids and other insect vectors are held in check by cool temperatures and strong winds. The environmental conditions for seed potato production in southern Chile are among the best in the world.

Most potatoes are planted in October and November and harvested in April and May. Farmers make heavy applications of livestock manure, but chemical fertilizers are not extensively used. On average, potato farmers apply about 200 kilograms of nitrogen per hectare, mainly in the form of livestock manure—nearly double the recommended rate. One reason farmers apply so much livestock manure is to improve soil structure and fertility for later crops in the rotation. Some farmers along the coast mix marine plants with the livestock manure they spread on the potato fields.

Utilization of the potato harvest is more diversified in southern Chile than in most other developing areas. About a third of the harvest is sold to urban food markets, 15 percent is shipped as seed to potato farmers in the Central Valley and other northern areas, and 10 percent goes to local starch factories. The balance is kept by farmers and used in rather equal proportions to plant in the next crop, to feed to livestock, and for family consumption. Most farmers store potatoes in multipurpose outbuildings. Thanks to the cool, damp weather after harvest, storage losses are generally less than 10 percent.

In recent years, prices for seed from southern Chile have been strong, but prices for processed forms of potatoes have been weak. Manufacturing of starch from potatoes is barely profitable. A small, modern plant for producing instant mashed potatoes opened in the 1960s, but it has not prospered because of the limited demand for convenience foods, competition from imported substitutes, and the irregular supply and price of potatoes available for processing.

Bibliographic notes

Few studies of potato production systems have been made in tropical and subtropical areas. In many places there is little accurate information on *where* potatoes are grown, much less *how.* Seminal references on farming systems, like Ruthenberg (1980), scarcely mention potatoes. For this reason, CIP has sponsored a number of studies of potato production systems in developing countries including Chile (Fu, 1979), Ecuador (Valderrama and Luzuriaga, 1980), Kenya (Durr and Lorenzl, 1980), Rwanda (Durr, 1980), India (Srivastava, 1980), Peru (Franco et al., 1979, 1980, 1981, 1983; Horton, 1984; Horton, et al., 1980; Mayer, 1979; Niñez, 1984; Vargas, 1983; Werge, 1977a, 1979a, b), and Sri Lanka (Rhoades, 1984). A set of "International Potato Reference Files" has been established at CIP as a repository for information on potato production and use in developing countries.

This chapter draws heavily on studies of agroecological zones and potato production systems done by Robert Rhoades, some results of which appear in Rhoades (1982a) and in CIP annual reports. This chapter has also benefited from many discussions with Efrain Franco and Enrique Mayer as well as the critical reviews of Richard Harwood and William O. Jones. The brief discussion of household gardens is based on Niñez (1984).

The Mantaro Valley case is based on Horton (1984); the Cañete Valley case, on Vargas (1983); the Philippine case, on Potts (1983); the Rwanda case, on Durr (1983); the Bangladesh case, on Elias and Islam (1982); the Chile case, on Fu (1979).

7 Research priorities and potato programs

Agricultural research and extension programs should respond to national development needs; in general terms, these are growth, equity, and food security. If a commodity research and development program fails to help to increase production, to narrow extremes in income inequality, or to assure a more stable food supply, there is little justification for its existence.

In nearly every developing country, agricultural growth—the rate at which farm production expands—must be accelerated in order to satisfy domestic needs without inflation or deepening dependence on imports of food and fiber. A lagging agricultural sector retards overall economic growth and welfare improvement. Most directly, it limits domestic food supplies and raises prices. It also slows gains in incomes and savings and the expansion of demand for agricultural inputs and consumer goods in rural areas. As food is a major expenditure for urban consumers in poor countries, rising food prices significantly reduce the amount of money they have for purchasing other things. Hence, through operation of what economists term "growth linkages" and "multiplier effects," slow agricultural growth has negative repercussions on incomes and demand throughout the economy. One of the most important potential contributions of a potato program to economic development is to expand food supplies and lower prices. For this reason, potato programs should select research and extension projects that promise high returns to investment through increased production or reduced post-harvest losses.

Policies and programs that accelerate agricultural growth do not always improve equity. But in the case of potatoes, rapid growth of output, expanded employment, and a more equal distribution of income often go hand in hand. To ensure this complementarity, potato programs should emphasize new technologies that employ the country's most abundant resources. Usually this means that new technologies should be labor-using and capital-saving.

The term *food security* covers two quite different concerns: the security of food supplies at the household level and the security of national food supplies and international availability. Recent instability of commodity markets and devastating famines in Africa have made food security a central issue for policymakers in many countries. One way to moderate fluctuations in food supplies, both at the household and national level, is to broaden the range of food crops grown. Potatoes can play an important role in diversifying cropping systems and food supplies in many areas.

Policymakers have a number of instruments at their disposal for achieving agricultural growth, improved equity, and food security. Potato research and extension programs warrant special attention from policymakers because (1) many requirements for success are different from those for programs in other crops, (2) the potential benefits of potato programs are greater than often realized, and (3) potato programs have received little attention to date.

The case for potato research and extension

Research and extension activities range from basic research to the diffusion of accumulated information. Compared with applied research and extension programs, basic research differs in three significant ways: It has less predictable outcomes; it requires more costly, specialized facilities and personnel; and it is more time-consuming.

Before 1900, only a few countries had developed the institutional capacity to generate and disseminate a stream of new agricultural technology. Several of the developed countries established publicly-funded agricultural research and extension systems in the early 1900s, but development of a similar capacity lagged far behind in the tropics and subtropics until the end of the colonial era. Research activities in developing areas centered on collection of exotic plants and improvement of export crops. Only since World War II has significant attention been directed toward the many crops and livestock species that are consumed in the tropics and subtropics.

Public support for agricultural research and extension is essential. Private firms cannot be relied on to conduct all of a nation's research because most of the benefits will accrue to other producers and consumers. As most improved agricultural practices and technological advances cannot readily be patented or marketed at a profit, private firms have little incentive to carry out agricultural research. In this sense, the benefits are "external" to the private firms' profit and loss calculations. (There are notable exceptions, of course, such as the development and marketing of hybrid maize and the successes of privately owned agrochemical and seed multiplication operations.) In contrast, firms in manufacturing and service industries can capture a large part of the benefits arising from their investment in research and development. They can "internalize" them.

The greater the "externalities" (the benefits that accrue to other members of society), the greater the case for public funding. The degree of externalities in commodity research is related to the type of new technology being developed, the elasticity of demand for the commodity in question, and the type of farmers growing the commodity. If the new technology is a product or process that can be patented and sold—farm machinery or hybrid seed, for example—private firms can be expected to carry out much of the research and development. But for research topics that will result in new information that cannot be profitably marketed, the public sector must step in.

If the demand for a commodity is elastic, most of the benefits of new cost-saving technologies are retained by producers, giving them a strong incentive to finance their own research and development. If, on the other hand, demand is inelastic, most benefits of technical change are passed on to consumers through lower prices, and producers have little reason to contribute funds for research and development.

Even if the research output is marketable and demand is highly elastic, it is unprofitable for small farmers to carry out research or extension programs unless they are organized to share in the costs and benefits. As efficient organization can seldom be achieved with large numbers of small farmers, producer groups that support research and extension programs usually involve a few large growers rather than many small ones.

The foregoing implies that to generate new information and technologies that cannot be profitably marketed, public support for research and extension activities is vital, especially when, as is usual with potatoes, the research is aimed at small farmers.

Setting research and extension priorities

A country's agricultural research and extension priorities should reflect its growth, equity, and food security goals. However, there is no simple, uncontroversial way to translate broad development objectives into specific priorities for research and extension. Ideally, the officials responsible for allocating public resources would have a clear set of objectives as well as procedures for estimating the potential contributions of various research and extension activities to each objective. They would also have at their disposal detailed information on the likely future streams of costs and benefits associated with alternative research and extension projects and use that information to select those projects with the highest social returns. However, conditions are always less than ideal and in most situations resources are allocated on the basis of much simpler approaches.

The parity or congruence model

The most widely used mechanism for assigning funds to commodity programs, or judging the adequacy of the present allocation, is known as the "parity" or "congruence" model. In this model the budgets for each program are compared with estimates of the economic value of the production of each commodity. Perfect congruence requires that the budgets for each program be in proportion to the value of production of the respective commodities. The usefulness of this model rests on two critical assumptions: that the opportunities for productive scientific effort and productivity-enhancing technical change are equivalent for each commodity, and that the value of scientific or technological advance is proportional to the value of production of the commodity. If these assumptions are valid, then allocation of resources using the congruence model will equate the rates of return to expenditures for each commodity and maximize net returns to agricultural research and extension.

But these assumptions cannot be valid because the opportunities for scientific advance and their social value are clearly higher for some commodities than others. The opportunities for applied potato research are higher than those of yams, cassava, and many other crops that are grown exclusively in the tropics and subtropics because national potato programs can tap the scientific resources of many well-established potato research programs in temperate-zone developed countries. The externalities associated with potato research and extension are also greater than those associated with many other commodities because few new potato technologies can be profitably marketed by private firms. As potatoes are perishable and bulky, and most are marketed domestically, the short-run price elasticity of demand is low. This provides little incentive for producers, traders, or processors to finance potato research and extension. Moreover, because most producers operate on a small scale and few producers' associations have been formed, it is difficult to marshall private resources for potato research and extension.

For these reasons, mechanical application of the congruence model is likely to result in underinvestment in potato research and extension. Nevertheless, congruence calculations can serve as a useful starting point in the resource allocation process.

In the funding of the research centers of the Consultative Group on International Agricultural Research (CGIAR), the proportions devoted to research on rice and wheat are significantly below their share in the total value of food crop production in developing countries. The proportions devoted to most other crops are slightly greater than their share in the value of food crop production (Table 17). But unlike CGIAR expenditures,

TABLE 17
Value of Ten Major Crops in Developing Countries Compared with Research Expenditures in the CGIAR System

Crop	1981/83 Commodity value (US$ billions)	(%)	CGIAR direct research expenditures (US$ millions)	(%)
Rice	68.7	45	23.6	36
Wheat	26.6	18	7.6	11
Maize	18.5	12	8.5	13
Potatoes	11.7	8	5.2	8
Cassava	8.9	6	6.4	10
Sorghum	5.8	4	3.2	5
Beans	4.3	3	6.2	9
Millet	3.9	3	3.0	4
Barley	3.1	2	1.7	3
Lentils	0.4	0	0.8	2

Source: Commodity value is based on production estimates for the period 1981/83 from FAO, *Production yearbook 1983* (Rome, 1984) and unpublished FAO estimates of average farm-gate prices for 1977 (the most recent date for which estimates are available) CGIAR expenditures are from FAO, Consultative Group on International Agricultural Research, Technical Advisory Committee, *TAC review of CGIAR priorities and future strategies* (Rome, 1985).

the proportion of national research budgets allocated to potatoes, other root crops, and pulses, is substantially below their share in the value of food crop production. This suggests many developing countries are underinvesting in research on these "minor" crops.

After examining initial congruence calculations, agricultural leaders should then establish the real determinants of payoff from research and extension—the costs, benefits, and externalities of research and extension on specific production and marketing problems. Because of the data and methodological limitations noted earlier, it is not possible to allocate resources on the basis of rigorous, comprehensive cost-benefit analyses. But an attempt should be made to identify the principal constraints to potato production and use and to assess, in qualitative terms (supported, wherever possible with quantitative evidence), the chances of removing these constraints, the cost of doing so, and the likely benefits.

Constraints to potato production and use

The importance of correctly assessing the constraints to potato production and use can scarcely be overstated. If a program is working on the wrong problem, no amount of research or extension can achieve the desired impact.

General constraints. Of the major constraints to potato consumption, high price is the one that can be most readily remedied by research and extension work. Hence, research and extension programs should focus on identifying and altering those factors that make potatoes expensive. Three means are available: lowering the prices of inputs, reducing profits at different stages of the production and marketing system, and improving the technologies employed by farmers and market agents.

Input price levels are, by and large, beyond the control of the research and extension system. However, researchers and extension workers should be aware of input prices and orient their work to generate and disseminate new technologies that use the most abundant and cheapest factors of production. Labor is often one of these.

Researchers can sometimes uncover monopolistic tendencies and excessively high profits within the marketing system. However, potato programs rarely have sufficient political influence to redress these problems.

A realistic goal for a potato program is to find ways to diminish the cost of producing and marketing potatoes by introducing improved technology. Production costs for potatoes exceed US$1,000 per hectare in most developing countries (as shown previously in Table 15). Only in areas like Rwanda, where farmers use few purchased inputs, are costs much lower. The single most costly factor of production usually is planting material, followed by labor and fertilizer. Expenditures on pesticides and equipment are generally less.

Because potatoes are grown from seed tubers, planting material is a much greater expense than in crops grown from true seed (such as cereals, pulses, oilseeds, and many vegetables) or from stem cuttings (such as sugarcane, sweet potatoes, and cassava). Planting material is most expensive in countries that lack seed production systems and must import seed tubers from abroad. In parts of Asia and Central America, seed tubers account for 40 to 50 percent of the total variable cost of production, whereas in much of South America and tropical Africa, they account for about 30 percent of the total. In extreme cases, such as the Philippines, farmers may spend over US$1,000 per hectare just for imported seed tubers. The high cost of seed tubers and their frequently poor quality are compelling reasons for potato research and extension programs to emphasize better seed potato production and distribution and to develop more cost-effective systems employing tissue culture, stem cuttings, or true potato seed.

The share of labor in total potato production costs tends to decline as economies develop and the relative importance of fertilizer, pesticides, and equipment grows. In Rwanda, for example, three-fifths of the production expenditure is for labor and almost none for purchased inputs. But in Mexico, labor constitutes a fifth of the production costs; fertilizer, pesticides, and equipment together account for over one-third. Often there is opportunity for cutting the labor input through mechanization. However, this would be inefficient because labor in developing countries is usually abundant relative to capital. Moreover, equity and food security goals are best met by creating more productive employment opportunities in agriculture. Thus, the emphasis should be on raising the productivity of labor in potato production by employing labor-intensive means of raising yields or cutting losses, such as better seed management, more careful control of pests and diseases, and improved post-harvest operations.

Pesticide use on potatoes has grown rapidly, particularly in areas with warm climates. Yet, pesticides still account for only a small share of total production costs, although the social costs are often high. Farmers are getting on a pesticide treadmill, which requires them to use more chemicals just to maintain the same or an inferior level of control. And many pesticides marketed in developing countries are highly toxic. For these reasons, potato programs need to investigate resistant varieties, biological control measures, and application of less toxic chemicals as ways to minimize pest damage.

Looking across the spectrum of inputs used in potato production, it is apparent that possibilities for lowering costs per ton of output are much greater than those for lowering costs per hectare. Developing new techniques and systems that decrease the amount of tubers that must be reserved as planting material is the principal area in which research can help trim production costs per hectare. It is unlikely that major savings can be made in fertilization. On the contrary, farmers will apply higher rates of fertilizer in the future. Ways may be found to use smaller amounts of costly pesticides, but these changes would have slight effect on total costs. Mechanization could shrink the labor input, but it is not likely to reduce total costs and probably would aggravate unemployment and social problems.

In most developing nations, the greatest payoffs from potato research and extension are likely to result from expanding potato cultivation, increasing yields, and reducing yield fluctuations and post-harvest losses. Achieving those objectives usually will require improvement in seed systems, varieties, pest management, and storage. Most potato programs should have some expertise in each of those four areas. Improvements in tillage, irrigation, fertilization, and other agronomic practices should have lower priority except in areas where the potato is a relatively new crop that farmers have not yet learned to manage well.

Specific constraints. It is dangerous to jump from a list of general problem areas to specific research and extension priorities, for this overlooks two further steps: identifying specific constraints within each area, and assessing the costs and chances of successful research and extension in each area.

Poor seed quality is a serious problem almost everywhere, but the specific difficulties to be overcome vary from place to place. Too often seed programs in developing countries focus on control of virus diseases, as is done in Europe. However, in developing areas, the poor physiological condition of seed tubers at planting time is more critical. For this reason, many attempts to transfer the European seed certification model to developing countries have failed, while activities aimed at solving storage and other physiology-related problems have flourished.

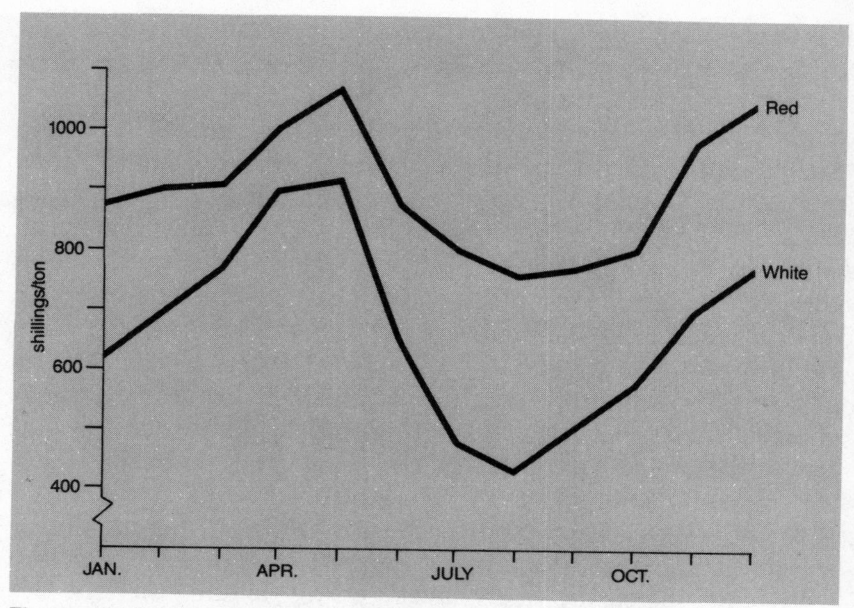

Figure 15. Comparison of average monthly wholesale prices for red and white potatoes in Nairobi, Kenya, 1973–77. *Source:* G. Durr and G. Lorenzl, *Potato production and utilization in Kenya* (Lima: International Potato Center, 1980).

In breeding or germplasm-screening programs, selection criteria should reflect not only the production problems as perceived by scientists, but also farmers' criteria for selecting new varieties as well. Breeding programs go astray when they focus on a single production problem to the exclusion of other important varietal characteristics, such as stability of yield, consumer preferences, and storability. In the 1960s and early 1970s, Kenya's breeding program, for example, identified new varieties that were highly resistant to late blight, the country's most serious potato disease. But few farmers adopted the varieties because they did not meet local market requirements. The new varieties were white-skinned, whereas consumers had a strong preference for red-skinned potatoes and were willing to pay a substantial premium for them (Figure 15). The yield advantage of the new varieties was not enough to compensate for their lower price. The breeding programs that have been most successful have maintained close links with farmers and have benefited from extensive on-farm testing of potential new varieties.

Priorities for pest management research should also reflect local conditions because the problems as well as the possible solutions may vary. In the long run, breeding for resistance is perhaps the most promising way to limit the economic damage from pests and diseases. However, where resistant varieties are not available, use of chemical pesticides is climbing. This trend reflects both the effectiveness of chemicals against many pests (at least in the short run) and the effectiveness of private companies in promoting their products. Pests can sometimes be controlled without or with fewer pesticides, but farmers may not be aware of alternatives. An important role for the public sector is to develop and promote alternative pest management techniques that do not rely on patented chemicals or that employ the safest ones. There is considerable scope for useful work in this area. In Tunisia, for example, the highly toxic insecticide parathion was widely used for control of tuber moth in potato storage until researchers informed farmers and policymakers of the dangers of this product to human health. Subsequently, parathion was banned and a less harmful pyrethroid—decis—has come into general use.

Among the many facets of post-harvest technology, seed storage merits the greatest attention in the short run. Consumer potato storage, however, usually receives more attention because of the erroneous assumption that price fluctuations are caused by inadequate storage. Attempts to make public agencies responsible for large-scale storage of consumer potatoes usually fail because the need for more storage is incorrectly assessed and the technical and managerial requirements of successful storage are underestimated. As most potatoes, both for seed and consumption, are stored by farmers, research and extension programs should focus on improving on-farm storage.

Specific storage designs are seldom applicable in different locations, but many storage principles are. Exposing seed potatoes to diffused light has aided seed storage in several countries. It seems likely that similar improvements could be made in on-farm storage of consumer potatoes. In some areas, research should also address off-farm storage. Cold storage of seed and, to a lesser extent, of consumer potatoes has revolutionized potato production in India and Bangladesh. The challenge now is to develop cheaper storage systems that are more widely applicable to the tropics and subtropics.

In most developing countries, there is not yet a sufficient demand for processed potato products to encourage much private processing research and development. The high cost of potatoes, lack of market infrastructure for handling frozen foods, and absence of a large restaurant trade in most developing areas make it unlikely that the demand for convenience foods such as chips and precooked and frozen french fries will expand dramatically in the foreseeable future. Moreover, where fresh potatoes are available in markets much of the year, it seems unlikely that the demand for dried flakes or granules will make inroads into the demand for fresh potatoes. All of this implies that before embarking on processing research, planners should carefully study the potential markets for processed potatoes.

Feasibility and costs of research and extension

After identifying the specific constraints to potato production and use and assessing the feasibility of research and extension projects, policymakers need to consider different lines of action in relation to feasibility—the probability of success in a given endeavor—and the costs—the necessary commitment of facilities, specialized personnel, operating funds, and time. Judgments on these matters should be formed with the help of scientists working in the specific areas concerned. As research is by nature a process of trial and error, precise figures cannot be expected. However, even broad ranges of magnitude are helpful.

Take, for example, a country in which potato production and use faces three major constraints: (1) the local variety is poorly adapted to the prevailing warm growing conditions; (2) imported seed is expensive and its quality is unreliable; and (3) existing cold storage capacity is inadequate. Scientists decide that the single technological improvement that would have the greatest impact on production and use is a new variety that is high yielding, that stores well under high temperatures, and that is resistant to virus diseases. Such a "miracle variety" could lead to a substantial gain in potato production without a seed certification program or the necessity for building additional cold storage capacity. Because no such variety exists, a breeding program is set up to produce one without regard for cost.

Under this scenario, all funds available for the potato program are invested in a breeding program that, after 20 years, fails to produce the desired variety. In response to the potato program's inability to generate a useful new technology, policymakers cut its funding and replace the management!

A more appropriate strategy might have been a four-pronged approach.

First, establish a low-budget screening program to select, from breeding materials available from CIP and other breeding programs, those varieties that are best suited to the country's needs. In screening, work closely with farmers and consider not only yield but marketability and storability. When short-term goals are met, consider a more ambitious breeding program.

Second, establish a seed program to import and multiply seed at least once before distribution to farmers. At the same time, explore the feasibility of replacing imported seed with locally produced seed. If this is not practical, contract with a foreign firm to produce a small amount of seed each year in the varieties identified by the screening program. Put initial emphasis on improving seed physiology rather than on controlling virus diseases. Explore the economic feasibility of new rapid multiplication techniques, including in vitro culture (test-tube multiplication of plants from tiny plant segments) and true potato seed, as they become available.

Third, promote expansion of cold storage capacity. Conduct storage research jointly with private firms. Explore possibilities for using evaporative coolers and other low-cost methods for storing potatoes. Give first priority to storage of seed rather than consumer potatoes. Look for cheap capital from foreign donor agencies to finance a cold storage structure and other "hardware" for the national research program.

Fourth, monitor successes and failures to gain experience for improving the program and to generate evidence needed to defend the program's budget in terms of its impact and social benefits.

This hypothetical case illustrates several significant points. First, it is important to be realistic in setting priorities and striking a balance between short- and long-term goals. While in the long run the miracle variety *may* have a high payoff, in the short run it has no payoff whatsoever. Certainly a potato program that does not generate technological improvements in the short run cannot expect much support for its long-term projects. Second, it is desirable to establish a diversified program with enough flexibility to take advantage of new opportunities, such as the new multiplication techniques, as they present themselves. Third, monitoring and evaluation is useful for the program managers as well as to demonstrate to outsiders the cost effectiveness of the program.

Successful programs attract additional resources, both from domestic and foreign sources. For this reason, potato program officials need to set realistic goals and solve significant problems facing producers, market agents, and consumers.

Potato programs

Assistance to potato programs

One can trace the beginning of international assistance in potato research in the Third World to at least 1943 when a joint crop improvement program between the Rockefeller Foundation and the Mexican government began. Late-blight resistant varieties developed under the program have been distributed to many other countries and are now widely grown in Central America, Rwanda, Nepal, and the Philippines. The Rockefeller Foundation also supported potato programs in other countries through financial aid, technical assistance, and training programs. Many former Rockefeller scholars now lead crop improvement programs in developing countries.

In the post-war period, the bilateral assistance agencies of such countries as Australia, Canada, Germany, the Netherlands, Switzerland, the United Kingdom, and the United States have helped establish or strengthen potato research and extension programs in a number of developing countries. Some of these, like the Dutch-supported program in Bangladesh, began with an emphasis on research, whereas others, like the U.S.-supported program in Peru, emphasized extension. Sooner or later, most programs have ended up stressing national seed multiplication systems.

Although bilateral support for potato programs has contributed significantly to potato research and extension in many developing countries, there have been shortcomings. Often the duration of support has been too brief. Many bilateral programs have sent individuals abroad for training, but by the time they returned, bilateral funding for the potato program might have been terminated. In large multicommodity programs potatoes have often been given a lower priority than cereal improvement. Another difficulty with bilateral programs was their lack of a solid research base in developing countries, and because there was no formal mechanism for communication among the bilaterally supported programs, researchers in different countries had little opportunity to learn from each other's experiences.

One of the reasons for creating the International Potato Center was to help surmount these problems by establishing a research program focused on potato production in developing countries and to link national programs with one another, with CIP, and with scientific and development agencies in developed nations.

The International Potato Center

Over the last two decades, agricultural research in the tropics and subtropics has received a significant impetus from the research centers sponsored by the Consultative Group on International Agricultural Research (CGIAR). The system of international agricultural research centers, established in 1971 with joint sponsorship of the World Bank, the United Nations Development Programme (UNDP), and FAO, is now supported by over 20 governments, international and regional organizations, and private foundations. It comprises 10 commodity-oriented research centers, of which the International Potato Center (CIP) is one, and three centers concerned with food policy, national agricultural research, and plant genetic resources.

Rationale for an international program. CIP was founded in 1971 as part of the CGIAR with the expectation that the payoff from international potato research would be high. A wealth of potato research results amassed in the developed countries was not being effectively used in many developing countries because they lacked strong potato research and extension programs. Bilateral programs provided only a partial solution to this problem. Where technological progress had been made, potato production was expanding rapidly, particularly in lowland areas. It was believed that an international potato research program could draw on the potato research already done in developed countries, concentrating its own research on topics inadequately covered by other institutions. Through applied research conducted jointly with bilateral agencies and developing country programs, it could overcome many of the constraints to potato production and use. Collaborative research, training, and improvements in the flow of information among potato scientists in different countries were expected to strengthen national potato research capacity, permitting a gradual transfer of responsibility from the international center to the developing countries themselves.

Objectives and priorities. CIP was established through an agreement with the government of Peru. CIP's headquarters are in La Molina, on the outskirts of Peru's capital, Lima, and it has experimental facilities on the coast, in the highlands, and in the Amazon basin. CIP's basic objectives are to enhance the potato's yielding ability, stability, and efficiency of production in developing areas and to improve the potato's adaptability to low tropical and subtropical regions as well as to cold, high regions.

Seven research priorities have been defined:

- Collection and classification of tuber-bearing *Solanum* species
- Breeding potatoes adapted to warm, low tropics and subtropics and to cool, mountainous environments
- Control of important pests and diseases
- Agronomy for warm areas
- Post-harvest technology

- Seed production for developing countries
- Socioeconomic studies

World Potato Collection. A process of "genetic erosion"—disappearance of primitive varieties and wild species of potatoes—is rapidly taking place in the potato's center of origin, the Andes. For this reason, an initial priority for CIP was assembling a collection of Andean cultivated and wild tuber-bearing *Solanum* species. Prior to CIP's establishment, only 10 percent of the known genetic variability of the potato had been used in potato breeding. For this reason, European and North American potato varieties have a narrow genetic background and limited disease and pest resistance. The World Potato Collection, grown each year at CIP's highland station in Peru, contains 5,000 cultivated and 1,500 wild accessions. Breeding material in the collection is available to breeders worldwide. At CIP it is being used to produce new genetic combinations with frost tolerance; adaptation to the lowland tropics; and resistance to viruses, bacterial wilt, the potato tuber moth, and nematodes.

Environmental adaptation. Most developing countries cultivate European and North American potato varieties that are not well adapted to their agricultural conditions. A goal of CIP's breeding program is to make available an array of breeding populations and potential new varieties from which national programs can select better-adapted varieties. A population breeding strategy is used to maintain wide genetic diversity while increasing the frequency of genes controlling adaptation, yield, and resistance or tolerance to biotic stresses. Research contracts link breeders at CIP headquarters with major potato breeding programs in Europe and in North and South America. Seventy potato programs in developing countries have received germplasm for evaluation. Thirty-six clones have been multiplied by 22 national programs and released to farmers. Nearly 350 clones now in advanced selection and variety trials in developing countries have resistance or tolerance to frost, nematodes, viruses, late blight, bacterial wilt, or heat.

Because of the need to meet stringent national quarantine standards, CIP has established regional germplasm redistribution centers in Kenya and the Philippines for testing, multiplying, and distributing selected genetic material to national programs.

The focus of breeding for lowland tropical conditions is on combining heat tolerance with earliness (60- to 75-day growing period), resistance to lowland pests and diseases (principally bacterial wilt, root-knot nematode, and potato viruses X and Y), and acceptable tuber quality (especially dry matter content, which is lowered by heat and high humidity). About 6,000 commercial varieties and breeding lines have been tested for adaptation to heat at a mid-elevation tropical site in eastern Peru, of which 34 were selected as heat tolerant and evaluated further at a lowland jungle site. Only four survived this rigorous test. Considerable progress has been made, but commercial varieties have not yet been released.

Two breeding populations have been developed for cool environments: one for the Andean highlands and the other for non-Andean cool areas. The national programs of Bolivia, Colombia, Ecuador, and Peru are testing the Andean populations; and Bhutan, Kenya, Nepal, Pakistan, and Papua New Guinea have received frost-resistant clones of the non-Andean population.

Pests and diseases. In pest research, CIP seeks to identify resistant varieties and management techniques for better control with more moderate use of insecticides. Among the pests and diseases of the potato, late blight, which is still the most widespread and damaging, has received special attention by CIP, as well as by other breeding programs in developing countries. In recent years, 36 varieties have been selected by national programs for late blight resistance.

No chemical control is available for bacterial wilt, which is a primary constraint to potato production in many lowland tropical areas. Following early work at the University of Wisconsin, CIP is breeding for resistance, and clones have shown resistance to bacterial wilt in Peru, Fiji, Sri Lanka, Nigeria, Nepal, Indonesia, Kenya, Egypt, and Brazil.

Some European varieties have cyst nematode resistance, but it is ineffective in the Andes. For this reason, one of the first programs CIP established was a search for sources of resistance to Andean nematodes. Biological control measures for the root-knot nematode are also being sought. A fungus that attacks the eggs of this nematode has been distributed to scientists for evaluation in 46 countries.

In entomological research, CIP focuses on control of the potato tuber moth, which causes heavy damage in the field and in storage in many developing areas. Twenty-two primitive varieties and 21 wild resistant accessions identified in the World Potato Collection are being used in breeding resistance to this pest. Techniques for using sex pheromones (attractants) have been developed for monitoring tuber moth populations as well as for direct control.

Breeders seek resistance to those viruses that are most damaging to potato crops in developing countries: potato leaf roll virus and potato viruses X and Y. Potato leaf roll virus and potato virus Y, which generally occur together, are the most serious, being easily transmitted by aphids. It has been possible to select materials resistant to leaf roll that are also immune to potato virus X and potato virus Y. Selection programs for virus resistance are active in a dozen developing countries.

Agronomy. Agronomic research at CIP is primarily aimed at developing practical recommendations for producing potatoes in warm areas. Results have pointed up the need to plant high-quality seed tubers in the correct physiological stage in order to get vigorous emergence and early ground cover. Intercropping to provide shade and use of mulch have proven helpful in lowering soil temperatures and boosting yields. Some intercrops and rotations have also decreased insect damage and incidence of bacterial wilt.

Post-harvest technology. CIP's initial post-harvest research centered on seed tuber storage. It was quickly found that excessive sprouting, high storage losses, and poor field performance could often be overcome by storing seed tubers in diffused natural light. Interdisciplinary research involving anthropologists and biological scientists determined that the principles could be applied in a number of countries. Seed programs in over 20 developing countries use diffused light storage, and farmers in nearly all those countries have adopted it. More recently, emphasis in post-harvest research has shifted to the control of pests in seed storage (principally aphids and the tuber moth) and to difficulties of storing and processing consumer potatoes.

Seed. Seed is the weakest link in most potato production systems in developing countries. In addition to breeding for virus resistance (aimed at retarding the spread of viruses through planting material) and work on seed tuber storage, CIP has been heavily engaged in training programs and technical assistance to strengthen national seed potato production and distribution systems.

Control of virus diseases, which is based on exclusion and prevention rather than cure, requires reliable methods of virus detection. Sophisticated detection methods have been simplified by CIP to better suit developing countries. A low-cost virus-testing kit has been disseminated.

A variety of practical techniques for rapid multiplication of virus-free planting material, including single node and sprout cuttings, stem cuttings, and leaf bud cuttings, have been developed and are routinely used by more than 20 national seed programs. Tissue culture methods are used in CIP's research, in disease-free maintenance of germplasm, and in the distribution of potato germplasm to other research programs throughout the world. As a result of training, a rapidly growing number of national programs are now able to apply tissue culture techniques and receive in vitro genetic materials from abroad.

Research on true potato seed was started in 1977 to develop an alternative to vegetative propagation of potatoes. True potato seed has potential applications in national seed programs as well as on farms. Its use could cut the cost of planting material and significantly increase the amount of tubers available for consumption. An estimated 10 million tons of potatoes are now planted annually in developing countries. True potato seed is much easier to store than seed tubers, and its use could greatly diminish the spread of tuber-transmitted virus diseases. Priorities for true potato seed research include developing uniform, early, high-yielding populations; improving seed production methods; and assessing the technical and economic feasibility of different multiplication systems using true potato seed in developing countries.

A handful of true potato seed is enough to plant a hectare of potatoes. One to two tons of tubers, sometimes more, are needed to plant the same field area.

Socioeconomic studies. When CIP was established, little documentation was available on potato production and use in developing countries. To fill the gap, anthropologists, economists, and sociologists were recruited to assemble and analyze information on potato production trends, farming systems, marketing, and consumption. Over time, the socioeconomic work has become increasingly integrated with biological research and has generated much new information on the role of potatoes in the food systems of developing countries, constraints to greater potato production and use, ways to improve the performance of potato programs, and the impact of successful programs. This information, much of which has been published in bibliographies, statistical handbooks, journal articles, and monographs, is intensively used at CIP as well as in national programs.

Regional programs. To ensure continuing scientific contact with national programs, CIP has established a network of regional offices—two in Latin America, three in Africa, and three in Asia. Regional scientists work with national researchers and extension workers in identifying production problems, testing new technologies, conducting adaptive research, multiplying and distributing genetic materials, and training potato workers. CIP's regional offices have no land or laboratories; they rely on national programs and in a few cases other international centers for their research facilities, office space, and logistical support.

CIP has formal memoranda of understanding with 44 research institutions in 33 developing countries. Special-project funding is occasionally used to post staff members in some of those countries, but the preferred relationship is assistance to national programs through collaborative research projects and contracts. Over a hundred collaborative research projects are conducted by CIP and research institutions in developing countries. Twenty-one research contracts facilitate research on priority topics for which adequate local funds cannot be obtained.

When national leaders seek external funding and technical support for a potato program, CIP attempts to link the program with a donor agency and a locally-based international agency to administer funds and implement the project. This arrangement, which allows CIP to collaborate with projects rather than manage them directly, is in effect in Bangladesh, Bolivia, Nepal, and Pakistan.

Training. Since 1978, more than 2,500 researchers and extension workers from 100 countries have participated in CIP training activities in the areas of germplasm management, production of true potato seed, potato production in warm environments, post-harvest technology, on-farm research, seed production (including pest management and tissue culture), and general production techniques. The goal is to improve the capacity of national programs to set appropriate research priorities, to carry out research in the most important areas, to disseminate research results, and to conduct their own training. Training plans reflect evolving research priorities as well as the capabilities and interests of national programs.

National potato programs

A decade ago, little potato research and extension work was being done in developing areas outside of South America, and

International training programs provide young scientists from around the world with up-to-date information on new techniques in potato breeding and related scientific fields.

research seldom was conducted within the framework of a national program. India and Mexico were the two major exceptions. Since then, due to a growing recognition of the value of the potato in developing countries and the stimulus and support provided by CIP and other institutions, research and extension activities have expanded greatly, and a number of coordinated national potato programs have been established in developing countries.

Potato programs vary greatly in their size, staffing, budgets, facilities, and organizational structures, but their major research and extension priorities are strikingly similar. Most programs place highest priority on seed production and selection of new varieties. Most also conduct work on pest management, post-harvest technology, and agronomy.

Potato breeding programs in nearly 20 countries are making their own crosses, and more than 70 are screening improved germplasm received from abroad. More than 40 developing countries have released new potato varieties in recent years.

Disease-free "mother plants" grown in greenhouses are the foundation of South Korea's seed potato certification program.

Forty-two national programs operate seed potato production and distribution systems. Twenty-three of these use stem cuttings and other rapid multiplication techniques, and 16 have tissue culture facilities for receiving in vitro materials and eliminating virus diseases. The main objectives of these programs are to control seed-borne diseases, introduce new varieties, and improve seed management. Thirty-six national programs are conducting research on true potato seed. Over 40 are conducting studies on seed-tuber storage, and 20 employ diffused light in their seed stores. Sixteen national programs are conducting research on consumer potato storage.

Although comprehensive information on pest and disease research is lacking, it is known that more than 20 programs are studying ways to improve control of the potato tuber moth. Many programs are also conducting research in the areas of mycology, bacteriology, nematology, and entomology.

Twenty-one developing countries are conducting agronomic research oriented toward the extension of potato production into warm areas; nine are investigating specific problems of potato production in cold areas.

Some developing countries that cannot afford a comprehensive potato program have formed collaborative research networks. CIP participates in five of these. In development of networks, collaborators identify major production constraints and country leaders assess the capacity of their own and other countries' programs to solve them. Responsibility for specific research projects is assigned to countries in accordance with their needs, interests, and facilities. This arrangement is especially useful for small countries with few potato growers, few scientists, and limited research facilities. Two factors essential for the success of research networks are strong coordination, particularly through the formative years, and external funding to start research projects and finance regional activities beyond the scope of national budgets.

Bibliographic notes

The economics of agricultural research has received much attention in recent years. Numerous technical papers on the allocation of resources in agricultural research are presented in conference proceedings edited by Fishel (1971) and Arndt, Dalrymple, and Ruttan (1977). Useful international comparisons are made by Boyce and Evenson (1975). Two more recent books, Ruttan (1982) and Pinstrup-Andersen (1982), are highly recommended as they cover the major issues involved in setting agricultural research priorities, are highly readable, provide up-to-date summaries of evidence on the subject, and present extensive references to more technical literature. Scobie (1984) provides a useful series of guidelines for managers of agricultural research programs and an extensive bibliography on the allocation of research funds with emphasis on agriculture.

The two sections of this chapter on research and extension draw heavily on Ruttan (1982) and Pinstrup-Andersen (1982). The third section on potato programs is based on International Potato Center (1984).

research programs and an extensive bibliography on the allocation of research funds with emphasis on agriculture.

The two sections of this chapter on research and extension draw heavily on Ruttan (1982) and Pinstrup-Andersen (1982). The third section on potato programs is based on International Potato Center (1984).

8
Impact of potato programs

In assessing the impact of agricultural research, it is useful to distinguish between two types of technology and their impacts. Production technology comprises the methods used to cultivate, harvest, store, process, handle, transport, and prepare food for consumption. Production impact is the physical, social, and economic effects of new production technology on crop and livestock production, marketing, and use, and on social welfare in general (including the effects on employment, nutrition, and income distribution).

Institutional technology refers to the methods that research and extension institutions use in generating and disseminating production technology. Institutional impact refers to the effects of new institutional technology on the capacity of research and extension institutions to generate and disseminate new production technology. Institutional technologies include procedures for genetic engineering, screening germplasm, disease identification and elimination, and rapid multiplication of vegetatively propagated crops, such as potatoes. They also include strategies for training, networking, conducting on-farm research, and planning and evaluating research programs. Institutional technologies are conventionally packaged in the form of research publications, trained personnel, and the recommendations of technical assistance missions.

The useful life of many production technologies is rather short, so a continual flow of new technologies is essential to maintain agricultural growth. For this reason, improvements in institutional performance may, in the long run, generate greater social returns than improvements in specific methods. Because of the location specificity of most production systems, institutions operating at the national and subnational levels have a comparative advantage in generating production technologies, whereas international programs have a comparative advantage in generating institutional technologies.

The first of the so-called international agricultural research centers produced new varieties of wheat and rice. However, air shipments of seed packets—the classical, physical technology transfer—are now only one of several mechanisms used by international centers to distribute their research output. And even the new seeds shipped by international centers are now best thought of as institutional technologies, instead of finished production technologies, as they are usually destined for breeding programs or germplasm evaluation trials rather than for immediate use by farmers. Perhaps the most important contribution of the international centers is to set the tone for agricultural research in developing countries. In this respect, numerous institutional strategies that were developed through the collaboration of international and national centers are now widely used in developing countries.

Methods for assessing impact

Production impacts

Estimates of the impact of new technologies on production are useful because they are the starting point for assessments of economic and social benefits at both the farm and national levels. Impact studies customarily employ a production economics framework to assess the contributions of research to changes in agricultural output. Requirements for the analysis include estimates of changes in production in regions or countries; information about major production systems, including old and new technologies, and the proportion of total output generated by each; and data on production functions for each system. If the effects of technological change on various segments of society are to be analyzed, estimates of supply and demand elasticities are also required. Unfortunately, these data are seldom readily available and often are only approximate.

Use of production economics to quantify the impacts of technological change is most straightforward if the physical environment (e.g., relief, soils, and weather) is uniform across all observation points, if the production process generates a single output and employs few variable inputs, if inputs and the output are easily quantifiable and uniform in quality, and if all inputs are purchased and the output is sold. When the number of variable inputs and outputs is large, their quality is variable, and production is partially or wholly subsistence-oriented, impact assessment using a production economics framework becomes complex; its results are influenced by a large number of simplifying assumptions of dubious validity.

Varieties vs. other technologies

Data on use of new and old cereal varieties, fertilizer levels, and yields are most readily available in Asia, where university training in production economics is also most advanced. Hence, it is not surprising that production function analysis has been used most widely for assessing the impact of new cereal varieties and associated inputs in Asia.

In several regions, increases in production and yields of crops like potatoes have been on par with or ahead of the increases for wheat and rice. In addition to improved potato varieties, technical advances have been achieved in pest control, seed systems, and post-harvest technology. But these changes have not been assessed, apparently because potatoes are considered to be unimportant, and the impact of these kinds of technological changes are less easily measured than the impact of a new variety alone.

Data on changes in pest management and the socioeconomic impact of the changes are difficult to obtain. Farmers' pest management strategies are complex, and their inputs and outputs vary qualitatively as well as quantitatively. Detailed understanding of the pests, farmers' control measures, and whether the impetus for change comes from public or private agencies or from neighbors is needed for reliably assessing the impact of government programs. Collecting this type of information would require intensive multidisciplinary farm-level research, which has rarely been done.

A new system for producing and distributing seed potatoes usually embodies a number of technical and institutional innovations and triggers a number of changes in potato production and use. These changes may affect the timing of planting, yield, length of the growing season, mix of crops in the farming system, potato prices, and per capita consumption. Assessment of the impact of a seed program requires estimates of difficult-to-obtain variables like the seed degeneration rate (the yield reduction over time that results from virus infection). An understanding of the institutions involved in seed production, inspection, testing, and distribution is also needed in order to estimate present and future costs.

Assessing the impact of new storage technology, especially for seed, is complex. An economic assessment of what might appear to be a simple improvement in seed storage has to incorporate a large number of qualitative changes in both inputs and outputs. This is because farmers seldom copy demonstration storage models. Instead they modify them to meet their own needs and budgets. Many farmers integrate improvements into existing storage sheds or rooms attached to their houses. In addition, the benefits may take many forms, including reduced sprouting, longer storage life, reduced storage losses, easier pest control, earlier field emergence and shorter growing season, higher yield, availability of seed for a new planting season, and higher selling price for seed tubers stored in light.

Assessing these types of impacts on potato production requires more fieldwork and a more flexible analytical approach than is needed for an economic assessment of the impact of a new variety.

Institutional impacts

International agricultural research centers supply national programs with institutional technology rather than production technology per se. They can make an institutional impact in the following ways:

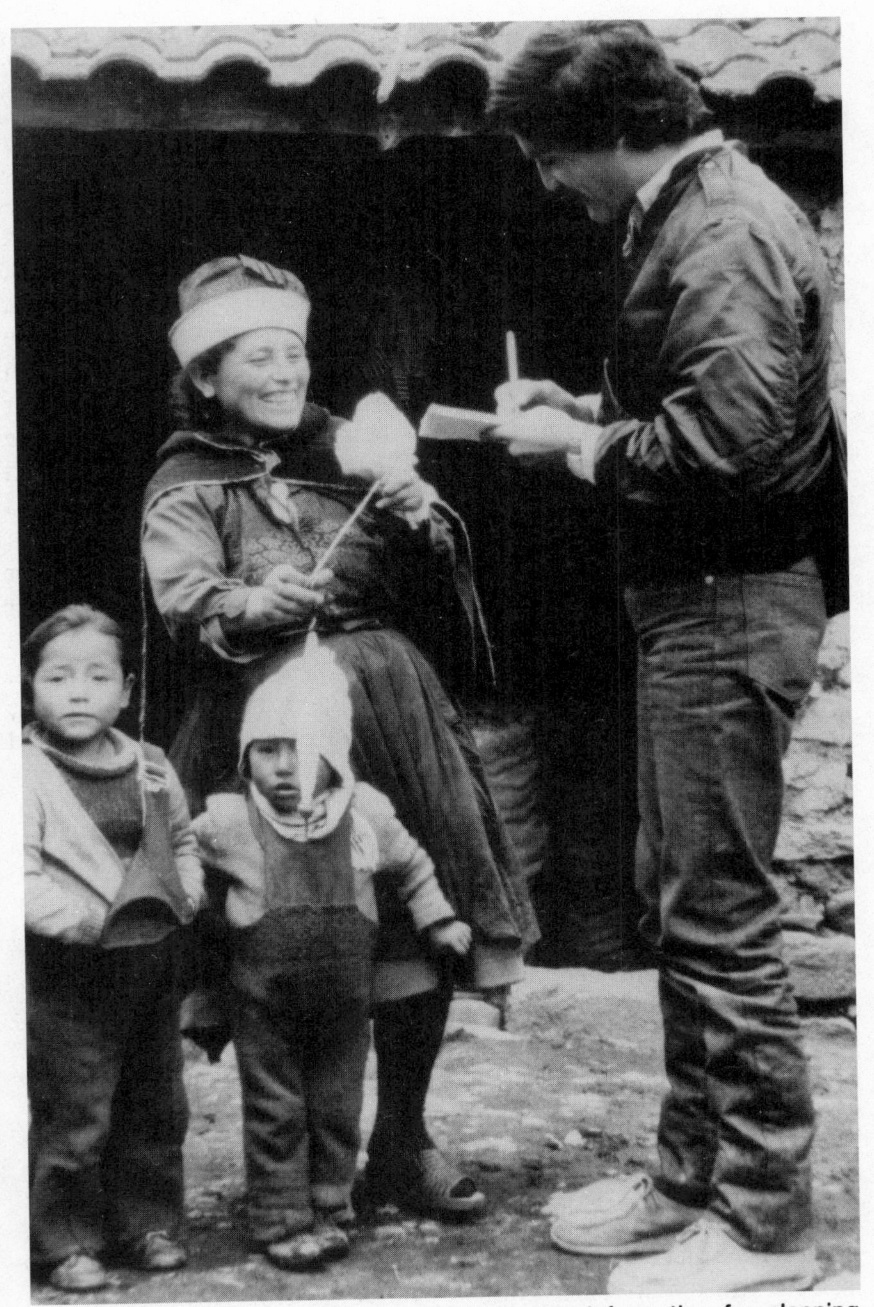

Discussions with farm families provide valuable information for planning agricultural research and development programs and for evaluating their results.

1. Strengthening agricultural research and development programs in developing countries, through training, supplying research information, etc.
2. Stimulating research institutions in developed countries to address problems that are important in developing countries
3. Linking national programs to the global research system, through improved communications, collaborative research, conferences, etc.
4. Contributing to establishment of priorities in key research and policy areas
5. Informing funding agencies and encouraging continued support from them

Given the diversity of potato programs and the many types of impact they have achieved, a case study approach is best suited to assessing impact. Instead of attempting to quantify the production impact of new potato technology at the farm level, the following brief case studies illustrate several potato programs and show how various types of benefits have been generated, ranging from training programs and institution building to yield increases and intensification of cropping systems.

Illustrative cases

Two criteria were used in selecting cases: (1) they should illustrate types of impact that could be achieved by other national programs, and (2) sufficient documentation should be available to describe the case. Several national programs met the first criterion, but had to be omitted because of the second criterion.

Expanding potato production on India's plains

For 300 years after their introduction to the Indian subcontinent, potatoes were mostly produced in the Himalayan hills. In the plains, potatoes were, until recently, confined to the fringes of towns, where they were grown for sale to the affluent. Large-scale commercial cultivation was not possible because of the inadequate seed supply and shortage of storage facilities. Seed for planting in the plains was imported from Europe for many years. In 1944, a seed potato production station was established at 2,400 meters above sea level near Simla in the northern hills, and as transportation improved within India, trade in seed potatoes between the hills and plains grew.

But due to the limited area suitable for potato cultivation in the Himalayas, the seed produced met only a small fraction of the potential seed requirement of the plains. Farmers in the plains could not plant an early fall crop because seed tubers were not available from the hills in time. Moreover, no potato variety could prosper unless it produced well under both the short-day conditions of the plains in winter and the long-day conditions of the hills in summer. To be successful, a variety also had to have a short period of dormancy, so that tubers harvested in the hills in September or October could be planted almost immediately on the plains. Phulwa was the only variety that met these conditions, and until recently, it was the most widely grown variety on the plains.

Most farmers in the plains could not keep their own potatoes for seed because their crops were heavily infected with virus diseases. Furthermore, high summer temperatures after the harvest made stored potatoes spoil rapidly.

As early as the 1930s, researchers found that populations of aphids, which spread virus diseases, normally were high for only 4 to 6 weeks, with the maximum infestation occurring in February. Later, based on this finding, researchers at the Central Potato Research Institute and its network of research centers developed a new seed production system for the plains, known as the seed-plot system. Under this system, farmers plant a small seed plot in early autumn and harvest it in January, before the buildup of aphids. The harvested seed is kept in cold storage until planting time the next season.

By following the recommendations, farmers can keep their own seed stocks in a healthy and highly productive state year after year. Seed produced via the seed-plot system is less costly than seed from the hills, and it is available in time to plant an autumn crop. It also has less late blight and bacterial wilt infection than seed from the hills.

Seed harvested on the plains and kept in cold storage over the summer emerges about 15 days earlier and produces an earlier harvest than seed from the hills planted at the same time. The main drawback of seed grown on the plains and kept in refrigerated storage is that it does not withstand rough handling and transportation as well as seed grown in the hills.

Once seed could be produced in the plains, it was possible for farmers to plant varieties that were adapted to the specific growing conditions of India's different regions. It also greatly simplified breeding work because it was no longer necessary to transport virus-free seedlings from the hills each year. Since the early 1960s, the Central Potato Research Institute has released a series of highly productive and disease-resistant varieties. Most, such as Kufri Sindhuri and Kufri Chandramukhi, were developed for use in the plains. Others, such as the late-blight resistant Kufri Jyoti, were developed for the hill areas. These new varieties have been rapidly adopted by farmers and are replacing Phulwa and other common varieties.

Development of the seed-plot system and use of the new Kufri series of potato varieties in the plains would not have been possible without cold storage. In the seed-plot system, seed tubers must be harvested before February and held for planting 9 months or more later in the next winter season. Not even Phulwa, which stores better than most new varieties, can be kept this long without refrigeration.

In 1955 the country's 83 refrigerated warehouses had a capacity of only 43,000 tons—an amount roughly equivalent to just 2 percent of the potato crop. Now there are about 3,000 refrigerated warehouses with a capacity of 4 million tons, of which 80 to 90 percent is used for potatoes. This represents about a third of the potato harvest. Most of India's refrigerated warehouses are privately owned, and the Central Potato Research Institute and collaborating institutions have conducted much of their post-harvest research in collaboration with private storage firms.

Because potatoes have a high water requirement, production in the plains has spread most rapidly in areas where canals and tube wells have been installed for irrigation. Aside from work on seed production, varietal development, and post-harvest technology, the Central Potato Research Institute and collaborating institutions have conducted important research on irrigation, fertilization, pest management, and cropping systems, which has

In India, rapid expansion of potato cultivation in the Punjab plains has resulted from introduction of new varieties, cold storage of seed tubers, and development of an effective seed program.

promoted rapid expansion in this nontraditional environment for the potato crop. As a result, potato production and consumption have increased sharply while potato prices have fallen. India is now the world's fifth largest potato-producing country, and the plains account for over nine-tenths of national production.

Rwanda's national potato program

Rwanda's national potato program, PNAP, is considered by many Rwandan authorities and foreign observers to be the country's most successful agricultural program.

PNAP was established in 1979 in cooperation with CIP. It was financed by the Belgian Technical Cooperation, the Rwandan government, CIP, and PNAP itself (from sales of seed potatoes). PNAP was given land for a 4-hectare research station and a 45-hectare seed farm. At present PNAP employs 5 Rwandan sci-

Preparing fields at the seed farm of the Rwanda national potato program.

entists, 2 expatriate CIP scientists, 10 technical and clerical workers, and about 80 field laborers. PNAP strives for simplicity and practical application of research results to solve farmers' problems. Its priorities are: selection of varieties that are resistant to late blight and bacterial wilt and acceptable to farmers and consumers; operation of a seed multiplication system; on-farm research; and training national researchers and extension workers.

More than 18,000 genotypes have been introduced and tested in PNAP's search for varieties resistant to late blight and bacterial wilt. Most were from the CIP, others came from Belgium and Uganda. PNAP selected and named four late-blight resistant varieties in 1982. Rapid selection of improved varieties was aided by weather conditions in the volcanic region, which permit two crops for selection each year; use of multiplication techniques (mainly sprout cuttings) to increase basic seed stocks; and on-farm trials, which incorporated farmers' views into the selection of potential new varieties.

Rather than attempting to operate an elaborate seed certification program, PNAP uses simple techniques to supply farmers with improved high quality seed. The seed production system is based on mass selection. In the absence of post-harvest virus-testing facilities, the system depends on field observation of plant vigor and the proportions of healthy and diseased plants. The program produces about 250 tons of seed per year, half of which is composed of new Rwandan varieties selected from genetic material introduced by CIP. Farmers are eager to replace their old varieties with new varieties resistant to late blight. Production of improved seed is still significantly short of farmer demand for seed, yet reports from seed projects and preliminary surveys indicate that about 7,000 hectares, or nearly 20 percent of the total potato area, are planted with seed originating from the national seed program. The average yield increase on farms due to use of improved seed is estimated to be about 3 tons per hectare—a 40 percent advantage over traditional seed.

After a short start-up period, the seed multiplication program's benefits have far exceeded its costs. An economic analysis indicates that as of 1985, the program's internal rate of return was estimated to be 40 percent—more than twice the return offered by most development projects in the country. The projected future rate of return is much higher because costs are expected to remain at about the same level while benefits increase significantly as the new varieties spread. Because virus diseases that depress yields spread slowly in Rwanda, the multiplier effect of a small stock of clean seed is great.

PNAP combines experimentation and technology transfer in farm trials. All PNAP scientists conduct on-farm trials as part of their research programs. One full-time scientist coordinates this work. More than 80 farm trials are conducted each year. The resulting close contact between researchers and farmers underlies PNAP's success in rapidly identifying production problems and providing farmers with appropriate solutions. During the first years of PNAP's operation, trials on farms helped determine the main constraints to potato production. Currently, these trials serve to screen and test all technologies developed by PNAP research or introduced from abroad.

Because of Rwanda's shortage of research and extension personnel, establishment of the potato program required training. As recently as 1982, the staff of the national institute of agricultural sciences included only one person who had obtained a Ph.D. in agricultural sciences. With CIP's help and through external funding, PNAP launched an intensive program of in-service short courses in Rwanda and abroad (Tunisia, Kenya, Peru, and the Netherlands). From 1978 to 1983, 23 Rwandan scientists and technicians participated in specialized courses and workshops. Two scientists obtained higher degrees. PNAP now holds two general production courses each year and a number of specialized training activities in which more than 200 local extension workers, students from agricultural schools, and farm managers have participated.

There are four keys to the progress of PNAP:

1. It has strong support from the Ministry of Agriculture; the government's financial contribution now amounts to more than 75 percent of PNAP's local operating budget.
2. It has incorporated farmers' points of view into research planning, implementation, and technology transfer.
3. It places priority on a few lines of action that address the most pressing needs of farmers and that do not require elaborate research facilities or complex institutional arrangements.
4. Its seed program is linked to a national seed service and development projects that distribute seed throughout the country, assuring rapid dissemination of improved varieties.

PNAP'S experience is being shared with neighboring countries through a regional network (Programme Regional D'Amélioration de la Pomme de Terre en Afrique Centrale).

Tunisia's seed program

Tunisia serves as an example of a small seed program that has generated an extraordinarily high rate of return. The government promotes cultivation of potatoes and other high-value vegetable crops in newly irrigated areas. Potato production has tripled since the early 1960s and now totals more than 140,000 tons from 12,000 hectares.

Potatoes are grown in three crops. The winter crop is produced only on the frost-free northern coast; spring and fall crops are grown there as well as in irrigated areas throughout the country. Tunisian farmers import European seed for planting the winter and spring crops. They plant the fall crop with tubers produced in the spring crop.

The Tunisian national potato program was established in 1976, with funding from the government and the Canadian International Development Agency. Under a cooperative agreement, CIP provided a full-time potato specialist as well as technical and administrative support. A technical committee was created within the Ministry of Agriculture to coordinate national policy for the potato crop. The planners and executors of the potato program worked to strengthen existing institutions and facilitate cooperative actions through the technical committee. This approach differs from that of many other development projects in Tunisia, which attempt to bypass local politics by setting up autonomous administrative structures.

Five Tunisian institutions are closely involved in the program. Groupement Interprofessionnel des Légumes (GIL) is responsible for seed multiplication; Centre de Perfectionnement et Recyclage Agricole (CPRA) is responsible for training extension workers and farmers; Institut National de Recherche Agronomique de Tunisie is responsible for research; Institut National Agronomique de Tunis (INAT) is responsible for degree training and research; and Direction de la Production Végétale is responsible for extension.

From the start, the potato program's leaders placed priority on the fall crop. Potato specialists had noted many single-stemmed plants in the fall season—an indication that at planting time the seed tubers were physiologically too young. Because potato yields are highly correlated with the number of main stems per plant, the specialists suggested that yields might be raised through simple management practices to increase the number of vigorous sprouts and stems produced by each seed piece.

GIL's seed multiplication specialists conducted on-farm trials to test the idea. The on-farm trials showed that yields in the fall crop could be nearly doubled through improvements in seed physiology resulting from:

- desprouting imported seed and early planting and harvesting of the seed crop in the early season
- culling of unhealthy seed tubers before storage (under straw) to avoid excessive losses
- desprouting locally produced seed tubers before planting in the late season

GIL exploited the research results in the seed multiplication system. CPRA and INAT incorporated findings into extension and training materials for farmers.

Beginning with 64 tons in 1977, the seed program increased production to over 2,000 tons per year in 1985. Yields on farms obtaining seed from GIL have increased by about 4 tons per hectare. The program's benefits have exceeded costs by a wide margin. Since 1976, the project's internal rate of return has been nearly 100 percent—a much higher rate than has been achieved in most other development projects in Tunisia or elsewhere.

Production of seed for the winter crop began in 1983. The goal is to replace imported seed with locally produced seed. The imported seed reaches Tunisia so soon after its harvest in Europe that it is still dormant and has a low-yield potential. On-farm trials indicate that seed produced in the spring and kept in cold storage from harvest (May-June) until planting time for the next winter crop (November-December) may yield as much as imported seed despite its higher levels of virus infection. More important, locally multiplied seed is cheaper than imported seed, is available for earlier plantings, and produces crops with a shorter growing period.

In addition to the "direct" benefits of the seed program—the yield increase on farms that obtain seed from GIL—there are substantial indirect benefits. The seed management practices recommended by the program have been adopted by many Tunisian farmers, increasing average yields in the winter crop by about 2 tons per hectare. The improved seed supply has also stimulated expanded planting of potatoes in the winter.

The key aspects of the success of the Tunisian seed program are: (1) it was established within existing national institutions and was perceived, from the outset, as a Tunisian program, although Canada provided funds and CIP provided technical assistance; (2) it used a problem-solving approach and focused on an apparently simple but important problem—seed in poor physiological condition at planting time; and (3) it ignored foreign models and was designed to meet local needs and economize on scarce capital resources and personnel.

Farmer use of tissue culture and rapid multiplication techniques in Vietnam

Tissue culture and associated methods for rapid clonal multiplication of plant tissues are often thought to be complex laboratory techniques with limited potential application in poor countries, particularly at the farm level. But Vietnamese farmers have refuted this idea. Since 1981, farmers in the Dalat area of Vietnam, with a minimum of facilities and training, have applied tissue culture techniques to produce potato planting material.

Potato production in Vietnam has expanded rapidly from less than 5,000 hectares prior to 1970 to 30,000 hectares in the mid-1980s. Annual production is about 250,000 tons. When imports of seed potatoes ended more than a decade ago, European varieties became increasingly infected with virus diseases. Therefore, the need to develop disease-free varieties became critical for the Vietnamese potato farmer.

In the 1970s, after the National Center for Scientific Research in Hanoi sent a scientist to France to learn clonal reproduction of plants, a tissue culture laboratory was established as part of the Center for Experimental Biology in Dalat. Tissue culture experiments began in 1977.

Technicians quickly found that a single in vitro cutting of a local potato variety, Thuong Tin (originally released in 1929 in Germany as Akersegen), could generate 4 or more cuttings a month. This implied that with subsequent multiplications, an annual output of more than 10 million plantlets was possible from a single in vitro cutting. They also found that these plantlets, when transplanted directly to the field, would produce several small tubers. Through trial and error, it was discovered that cuttings could be multiplied in nonsterile subsoil mixed with livestock manure, eliminating the need for costly growth sub-

Seed programs in more than 20 developing countries are using in vitro propagation of potatoes.

stances, sugar and agar. These findings, coupled with strong farmer demand for better planting material, encouraged the research station to develop a multiplication system for potatoes based on tissue culture and stem cutting techniques.

In 1979 the Vietnamese scientists began testing improved germplasm obtained from CIP. The research station at Dalat received 17 in vitro varieties that were adapted to Dalat's short-day conditions and could be grown year-round in the area. Some were more resistant to late blight than the European varieties previously grown. From that germplasm, three new varieties were released.

Researchers also developed techniques to extend the life of mother plants thus increasing the number of cuttings that could be harvested from each one. To prevent high mortality during the dry season, they rooted cuttings in pots that could be protected from sun and wind. Of several types tested, banana-leaf pots were selected because they were inexpensive, easy to make, and held soil moisture well.

The first potato crop produced from cuttings of the three new varieties was harvested in 1980. The new varieties averaged more than a kilogram of tubers per plant from transplanted cuttings and had almost no late blight infection. Farmers in Dalat began to buy rooted cuttings from the research station, and then, as demand exceeded the supply, several farmers began to root cuttings themselves, following the steps they had observed at the station. To reduce the number of cuttings they had to purchase, some farmers began to use the apical cuttings as mother plants. Eventually a few farmers began to multiply plantlets in vitro. The most successful farmers developed simple homemade laboratories and rooting areas for rapid multiplication in their gardens. These farmers, using mostly homemade equipment, recovered their investments within a few months.

During 1981, 10 farmers established rapid multiplication centers and sold rooted cuttings (transplants) to other growers. Annual production of transplants reached 2.5 million a year in 1982 as a result of the strong demand for the three new varieties. In 3 years, all the potatoes grown in Dalat changed from locally degenerated stocks to high-quality seed of the three new varieties.

As first- and second-generation seed tubers of these varieties became available, sales of transplants shrank because farmers found seed tubers more convenient to handle and easier to plant. The price of transplants has fallen to about the same level as a first-generation tuber.

Nevertheless, a market for transplants still exists. Virus incidence inevitably builds up and lowers potato yields in the third generation after a cutting-produced, virus-free crop has been planted, so farmers devote about 5 percent of their crop area to transplants to provide sufficient first- and second-generation seed tubers for the rest of their production.

In 1985 there were three family operations producing about 600,000 cuttings a year, which were purchased and planted by approximately 1,200 farmers. They have established reputations for providing high quality transplants. The largest one guarantees quality by allowing buyers to pay a month after transplanting for only the cuttings that survive.

This example demonstrates that farmers can use tissue culture to propagate potatoes. The technique does not require costly

In Dalat, Vietnam, farmers grow potato seedlings for sale to neighbors as planting materials.

laboratory equipment and highly trained technicians: Vietnamese farmers built suitable facilities and operated them successfully. The approach could be duplicated elsewhere in the developing world.

Improving storage in Sri Lanka

In Sri Lanka, rapid improvements in seed potato storage have had far-reaching effects not only on potato production, but on cropping patterns in market-gardening areas, as well.

In 1967, the government banned imports of consumer potatoes to save foreign exchange. As imports had been the bulk of the nation's supply, prices shot up, and vegetable growers in the highlands began expanding potato production. Potatoes have two main seasons. The early season crop is grown at lower elevations of the highlands from mid-May to September. The late-season crop is grown at higher elevations from October to mid-January.

For growers, the scarcity and high cost of seed tubers, most of which were imported, were severe constraints. Most seed for the early crop came from Australia and seed for the late crop from Europe. In 1979, imports from Australia were banned. The resulting shortage of seed for the early crop motivated farmers growing the late crop to begin supplying seed. However, they had difficulty storing seed from January to May when the early crop is planted. At the request of the Department of Agriculture, CIP began collaborating with potato specialists in Sri Lanka to solve the seed storage problem.

At a CIP-sponsored storage course in the Philippines, Sri Lankans were introduced to the principle of diffused-light seed storage, which retards sprouting and reduces storage losses. They also visited highland areas in the Philippines where diffused light storage was being rapidly adopted by farmers. Upon their return to Sri Lanka, experimental diffused-light storage structures were set up on government farms and a few private farms. Short courses on seed production and storage methods were also held for farmers.

Farmers quickly recognized that storing seed potatoes in diffused light not only reduces sprouting, which affects field emergence and stem density, but spoilage and infestations of insects like tuber moths can be more easily detected and remedied. By 1983 more than 500 farmers had built diffused-light storage structures and perhaps a thousand more had altered existing storage structures to allow the introduction of light. Even after adopting the new seed storage technology, two-thirds of the farmers have further modified diffused-light seed potato storage techniques to fit their needs.

As a result of adopting the new seed storage technology, farmers have increased their yields, they have adjusted their cropping patterns to include more potatoes, and they have increased their incomes. The hardships of acquiring seed from outside sources have eased, and the time allocated to storage activities has dropped.

Farmers have found that they can hold seed potatoes much longer than was possible in the traditional dark storage and that

Diffused-light storage is being used by increasing numbers of farmers for holding seed tubers from harvest to planting.

they can get higher prices for seed tubers stored in light. As a result, some farmers in upland areas who harvest in the late season keep seed for 6 months or more until the next late-season planting. And some farmers with well-drained paddy land in lower areas are storing seed from their early harvest until planting time the next season. Previously, farmers could only plant when imported seed was available from Europe or Australia, but now use of diffused-light storage has freed farmers to plant potatoes whenever growing conditions and market prices are favorable.

Increased local seed production and storage is allowing the government to reduce seed imports and save foreign exchange. The reputation of Sri Lanka's extension service has also been enhanced by the successful introduction of diffused-light storage.

A research network in Central America

Many developing countries are too small to mount a comprehensive potato program. For them, a practical solution is participation in a regional network in which several countries share the costs and benefits of research. One such network is PRECODEPA (Programa Regional Cooperativo de Papa).

Fifteen countries in Central America and the Caribbean produce potatoes, but, with the exception of Mexico, the average national production is only about 30,000 tons. Clearly, none of the individual countries can afford a comprehensive potato program. Moreover, it is difficult for CIP to justify intensive interaction with scientists in each country because the potential benefits are small relative to the potential in larger countries in other regions. Nevertheless, because potato production is expanding rapidly in Central America and the Caribbean—much more rapidly than in the rest of Latin America—there is a growing demand for new potato technology.

PRECODEPA was established to allow each country to specialize in the research areas in which it had a comparative advantage and to trade research results with others. Additionally, the network provides a mechanism for CIP personnel to work effectively with specialists in the region.

In 1978, representatives of Costa Rica, the Dominican Republic, Guatemala, Honduras, Mexico, and Panama met to discuss creation of a collaborative research network. The meeting reviewed the main factors limiting potato production and use in the region. Participants outlined their own country's problems and then drew up a list of problems that were common to several countries. The next step was to analyze the research interests and capabilities of each country. Finally, specific research responsibilities were assigned to the various national programs. CIP was asked to provide training and technical assistance in certain areas, such as breeding for resistance to bacterial wilt.

PRECODEPA began operations in 1978 under an agreement signed by the representatives of the six countries and CIP to implement the collaborative network. The Swiss Development Cooperation (SDC) provided funding for the network's research, training, consultations, and coordination activities.

The ultimate authority of PRECODEPA is the Permanent Regional Committee composed of the national coordinator of potato research and the other directors of agricultural research from each country. CIP also has two representatives. At its annual meetings, the committee evaluates and approves members' budgets and work plans and makes related policy decisions. An executive committee, consisting of a coordinator and two other country representatives, is in charge of project execution. This committee meets three or four times each year.

At the request of the Permanent Regional Committee and the SDC, CIP handles the disbursement of funds according to the approved budget, provides accounting, and arranges for external auditing. For the first 2 years, CIP provided the network's coordinator. Since then, the permanent committee has appointed a coordinator from among the participating national scientists. Periodically the coordinator arranges for an external review of each research project. In 1984 an external review of the entire network was conducted.

Nine projects were established in 1978. Since that time, the country leadership of two projects—bacterial wilt and socioeconomics—has changed. Three projects—virology, storage, and early blight—have been terminated, and a project on processing has been added. In recent years, El Salvador, Haiti, Nicaragua, and Cuba joined the network.

Originally, each country concentrated almost exclusively on its assigned research projects. But as local expertise has grown, countries have broadened their research, receiving advice and support from specialists in the leader countries. Progress has been most significant in the areas of seed production, late blight resistance, and seed storage techniques. Recently, country evaluations on control of the potato tuber moth and the golden nematode have begun.

Improved seed production systems, late-blight-resistant varieties, and simple storage structures are all having farm-level impact. All six original member countries are using seed production schemes. Farmers in five countries are growing Tollocan, a blight-resistant Mexican variety. Diffused-light seed storage is extensively used by farmers in Guatemala, and this technology has contributed to the successful establishment of Honduras' seed program.

A key factor in institutional development is training. Between 1978 and 1984, PRECODEPA organized 14 regional seminars, workshops, and production courses. The participation of 110 individuals in these events strengthened the technical expertise of member countries. An additional 22 national courses for technicians, extension workers, and farmers attracted 470 participants. CIP provided individual training for 20 scientists.

Certain intangible results of PRECODEPA are important, too. For example, personnel turnover in the member countries' potato programs is now lower than before, and lower than in other commodity programs of the region. The 1984 external review concluded that by providing members with expanded mutual support, expanded research responsibility, and frequent participation in regional meetings and courses (both as trainees and instructors), PRECODEPA has raised their level of job satisfaction. Another significant turn of events is the increase in the number of staff positions for potato research and development that each country provides from its national budget.

Lessons from the cases

Several lessons for designing and implementing potato programs can be derived from these cases.

1. Strong national commitment is essential. In every case, national policymakers, researchers, or extension workers provided leadership and financial support for the potato program, rather than depend entirely on a foreign agency. In India, Vietnam, and Sri Lanka, the programs were staffed and funded locally from the outset. They occasionally sought training, technical assistance, and financial support from abroad, but at no time did a foreign agency provide leadership or resident personnel for the program. In the cases of Rwanda, Tunisia, and PRECODEPA, foreign agencies played important roles initially in catalyzing and supporting local efforts. Foreign agencies provided a coleader for PNAP in Rwanda, a resident scientist in Tunisia, and a coordinator for PRECODEPA in Central America. But in each program, local leadership quickly emerged, and within a few years the budgets were financed principally from national resources.

2. Potato programs must blend into local institutions. The organizational structures of successful potato programs vary greatly. For example, India has one of the world's largest and most complete potato programs; PNAP in Rwanda is much smaller and less comprehensive; Tunisia's seed program involves five public institutions coordinated by a technical committee; the Dalat seed program in Vietnam involves a single institution working with farmers. Although none of these organizational arrangements are necessarily ideal, each has proven to be highly effective. Building the potato program within existing local institutions is essential to ensure its acceptability and continuity. Many programs have failed because they attempted to bypass local institutions rather than work within them.

3. Successful programs have clear priorities. Few countries can afford a comprehensive potato research and extension program. The cases show that a potato program can make a significant impact by addressing just a few key constraints to potato production and use, setting aside other problems for a time. For example, the careful research on aphid populations in India, which led to development of the seed-plot production method, opened the way for immense expansion of potato production and consumption on the Indian plains. Dissemination of improved storage methods has allowed Sri Lankan farmers to intensify their market-gardening systems by planting potatoes when imported seed tubers are not available. In Central America, each of the PRECODEPA countries is now focusing on a few key production problems and sharing information with the other member countries.

4. *Most successful potato programs have paid close attention to seed production and distribution.* Whereas improvement programs for other crops generally start with the introduction of new varieties, the most common theme of successful potato programs is seed improvement. In every case reviewed, potato programs placed high priority on improving the seed supply. Due to their bulkiness, perishability, high cost, and variable quality, seed tubers are a critical element in potato production systems, and better seed can benefit producers and consumers in many ways. In India, development of the seed-plot system allowed plains farmers to break their dependency on the limited supply of seed from the hills. Similarly, improved seed storage in Sri Lanka is allowing farmers to spend less on imported seed. In Rwanda and Vietnam, new seed systems have allowed the rapid evaluation and dissemination of better potato varieties. These and other experiences suggest that in most developing areas, the first priority for a potato program should be to identify and solve the major constraints in local seed systems.

5. *Small programs can generate high returns.* Many seed programs in developing countries attempt to supply a large proportion of all the seed potatoes required by farmers, but few succeed. Most produce no more than 5 percent of the seed required, and few produce as much as 20 percent of the total requirement. The inability of a program to meet its seed production goal often results from an overly optimistic forecast rather than from poor performance. Farmers in many areas can multiply their own seed two or three times, and sometimes more, before virus infection substantially reduces yields. Hence, a reasonable target for many seed potato programs is no more than a third or a fourth of farmers' total seed requirements.

As the cases have shown, a small, well-managed seed program can generate high returns to expenditures. Making available a small quantity of high-quality seed on a timely basis can have a large multiplier effect. Rather than being concerned with the volume of a seed potato program, managers should focus on the quality of the seed produced and the rate of return generated by the program.

6. *Technology cannot be directly transferred; it must be adapted to local conditions.* In none of the cases examined was foreign technology applied directly by researchers or farmers. Local adaptations were needed. The commercial application of tissue culture methods to seed potato production in Vietnam, establishment of an economically viable seed multiplication system for Tunisia's fall crop, and the rich variation of practical applications of the diffused-light seed storage principle in several developing countries provide striking illustrations of this point.

Bibliographic notes

Ruttan (1982) and Pinstrup-Andersen (1982) provide summary statements of the extensive literature on assessment of the impact of agricultural research. Scobie (1979) makes a more extensive review of evidence from developing countries. CGIAR (1985) summarizes results of a comprehensive study of the achievements and potential of the international agricultural research centers.

The sections of this chapter on types of impact and methods for assessing impact are based on Horton (1986). The illustrative cases are drawn from Pushkarnath (1976) for India; Monares (1984) for Rwanda; Horton et al. (in press) for Tunisia; Uyen and van der Zaag (1983, 1985) for Vietnam; Rhoades (1985) and Somaratne (1985) for Sri Lanka; and an unpublished text prepared by Kenneth Brown for Central America. Detailed versions of several of the cases appear in International Potato Center (1984).

9
Summing up

The agricultural agenda for developing countries is clear cut. They must accelerate their agricultural growth by intensifying their farming systems to feed a growing population, to meet future demands for livestock feed and fiber, to employ a growing labor force, and to reduce dependence on foreign supplies of agricultural commodities. The potato can play a useful role in many countries' agricultural development strategies. This chapter shows how, by providing answers to the 11 questions posed in the introduction to this book.

How important are potatoes in developing countries, and what are the recent trends in potato production and use?

The potato is often thought to be a crop of the industrial nations and of minor importance in developing areas. Developing countries, however, now produce a third of the world's potatoes. Since 1950, potato yields have doubled in developing countries and production has tripled. The growth of potato production, which exceeds that of most other food crops, is particularly rapid in Africa, Asia, Central America, and the Caribbean. In monetary terms, potatoes are now the fourth most important food crop in developing countries, after rice, wheat, and maize.

Despite the rapid changes, potato production still averages less than 30 kilograms per head in developing countries, of which two-thirds is for human consumption. Livestock feed and industrial uses together absorb only about 15 percent of production, and the remainder is used as seed or wasted. Because potato consumption still averages less than 20 kilograms per head in most developing countries, there is considerable room for increased consumption. For example, Western Europe consumes an average of 80 kilograms per head.

Contrary to popular perception, potato production is also increasing in most industrial areas. Since 1950, it has increased by about 80 percent in Australia and New Zealand and by more than 40 percent in the United States and Canada; it has remained about constant in Eastern Europe and the USSR. The only region where potato production has fallen significantly is Western Europe. This is because changes in farm organization, income growth, and rising potato prices have caused historically high levels of per capita production and consumption to fall during the last half century.

What biological or physical features make potatoes special?

Potatoes, like other root crops, generally yield more food energy per hectare than cereals. Due to their high protein-to-calorie ratio and short vegetative cycle, potatoes yield substantially more edible energy and protein per hectare and per day than both the cereals and cassava. The potato crop's high yield per unit of land area and time is an especially valuable trait in developing areas where the climate permits more than one crop to be grown in the field each year. Due to their high water content (about 80 percent), tubers are bulkier and more perishable than grains (which generally have less than 15 percent moisture). This makes potato storage, processing, transportation, and other post-harvest operations costly in the tropics where hot weather, insects, fungi, and bacteria can result in severe post-harvest losses. For this reason, post-harvest technology is a vital area for potato research in developing countries.

As potatoes do not form tubers unless average night temperatures fall below about 20 degrees C, they are not an economic crop in areas with high temperatures throughout the year. They are grown, however, in many lowland tropical areas that have 3 or 4 months with moderate night temperatures.

With 8 cultivated species and over 200 wild ancestors, the potato is genetically one of the most complex and diverse of all food crops. None of the major cereals, pulses, or other root crops approaches the genetic diversity of the potato. This is one of the potato's greatest assets, allowing it to be grown under a wider range of environmental conditions than most crops.

Among the Andean varieties and their wild ancestors can be found resistance to many pests and diseases, adaptation to extremes of cold and heat, and a vast array of tuber colors, sizes, shapes, and compositions. Most potatoes grown outside of South America have been derived from the few potatoes taken to Europe following the Spanish conquest. Most European varieties tend to perform poorly in the tropics, and, increasingly, scientists are turning to Andean potatoes as a source of desirable traits for breeding. The release of potato varieties adapted to tropical growing conditions and resistant to late blight and other pests and diseases has helped to increase yields, expand the potato producing area, and reduce production costs per kilogram.

Due to the cultivated potato's broad genetic background, the offspring resulting from sexual reproduction are highly varied. Uniformity and varietal purity can only be maintained by planting tubers or other parts of the stem. Farmers generally plant 1 to 2 tons of seed tubers per hectare. The large volume of seed tubers that has to be produced, harvested, handled, stored, hauled, and often desprouted before planting in the next season makes potato production expensive. The use of tubers also limits where and when potatoes can be planted because the maturity and condition of seed tubers influence the emergence, vigor, and yield of subsequent crops. Tubers can also spread pests (like nematodes) and diseases (viruses and bacteria) that depress yields and reduce tuber quality.

The limited supply and high cost of seed tubers are the major constraints to potato production in many developing areas. The difficulties of producing and distributing high-quality planting material for potatoes are much greater than for the cereals, especially in warm areas where virus diseases spread rapidly and storage costs are high. For this reason, potato programs need to pay special attention to improving seed systems for lowland tropical areas.

The potato is highly responsive to improved tillage, irrigation, fertilization, pest control, and management. Per hectare, most farmers invest more in potatoes than in their other crops. That investment is threatened by climatic hazards, pests, and diseases that can sharply lower yields. Two important goals for potato research and extension are to reduce the costs and risks of potato production.

Because potatoes are not easily stored for long periods or easily shipped over long distances, fluctuations in supply may result in sharp price changes, which compound production risks with price risks. Public agencies can help farmers cope with supply and price fluctuations by gathering and promptly disseminating information on areas sown, expected yields, and market conditions.

Who grows potatoes in developing countries, where, and how?

Some of the world's largest and smallest, richest and poorest, most progressive and most backward farmers grow potatoes. Most potato producers can be categorized as subsistence farmers, commercial farmers, or market-gardeners.

Subsistence farmers tend to have smallholdings and manage highly diversified systems with heavy labor use. They purchase few inputs and get low yields. Commercial farmers generally have larger and more specialized farming operations; most use chemical fertilizers, pesticides, and machinery; and they harvest higher yields. Market-gardeners differ from both these groups in that most have little land, but they are commercially oriented. They generally make intensive use of both labor and purchased inputs, and they achieve good yields. Aside from these three main types of potato farmers, uncounted households, both rural and urban, grow potatoes, along with other vegetables, in home or kitchen gardens.

Subsistence potato growers in isolated mountain areas in Rwanda and Bolivia are among the world's poorest farmers. Commercial growers in northern Mexico and Brazil are among the richest. Market-gardeners in parts of Guatemala and the Philippines are among the world's most intensive and productive.

Outside of subsistence areas, most potato producers are among the better-off farmers in their areas. The high returns to potato cultivation have stimulated rapid increases in potato production in developing countries.

Potatoes are grown under a wider range of altitude, latitude, and climatic conditions than any other major food crop. They are grown from sea level to over 4,000 meters elevation and from the equator to more than 40 degrees north and south. The diversity of agroecological zones in which potatoes are grown practically defies classification. However, three extreme types of production zones can be identified:

- *Highland tropical zones* such as the Andes; the Himalayas; and mountainous areas scattered throughout Africa, Asia, Central America, and Oceania
- *Temperate zones* such as southern Argentina and Chile, the Korean peninsula, northern Turkey, and northern China
- *Lowland tropical zones* such as the Indo-Gangetic plain from Pakistan through India into Bangladesh, Peru's coast, and northern Mexico

Highland farming systems vary greatly from place to place depending upon markets and environmental conditions (elevation, latitude, topography, soils, rainfall). In cool high areas, most farmers plant potatoes in spring and harvest them in the fall. At intermediate elevations, potatoes may be planted in two or more growing seasons each year. Subsistence farmers in highland areas usually plant several varieties in order to minimize production risks and vary their diets with different types of potatoes. They also purchase few inputs, harvest relatively low yields, and keep most of their potatoes for home consumption and seed. Market-gardeners and large-scale commercial potato producers in highland areas usually grow fewer varieties, purchase more inputs, harvest better yields, and sell most of their output. In most highland areas, mechanization is limited by the uneven topography and the small size of fields. High production risks discourage farmers from using credit.

In temperate zones, most potatoes are planted in spring and harvested in the fall. Cold winter weather facilitates storing potatoes for sale later in the year, for home consumption, and for planting in the next season. Use of machinery and chemical fertilizers varies greatly from place to place, depending on the degree of market integration, relative costs, rotations, and other aspects of the farming system. Insecticide use is generally lower than in lowland tropical areas, except on seed crops. Fungicides are often used to control late blight.

In many highland and temperate zones, climatic hazards like frost, hail, and drought are major sources of risk and depress yields. Late blight affects most highland and temperate areas; bacterial wilt is common in mid- and low elevations; insects and nematodes pose problems in some places. Because many small farmers cannot afford manufactured inputs, new varieties that are resistant to climatic hazards, pests, and diseases can have a major impact both on production and social welfare.

In the lowland tropics, farmers generally plant potatoes at the beginning of the cool season and harvest them at the end. Most lowland potato-growing areas are irrigated or have abundant residual soil mosture for winter production. Farming systems do not vary as much as in highland areas. Most farmers are commercially oriented, and plant only one or two varieties of potatoes; many use credit to purchase inputs. In many places high summer temperatures after harvest make potato storage difficult. Costly refrigerated warehouses are used to store both seed and consumer potatoes.

In the lowland tropics, potatoes are subject to fewer climatic hazards but more pests and diseases than in cooler areas. Because of the high ambient temperatures and insect populations, it is difficult for lowland farmers to produce and store high-quality seed tubers, and many buy seed produced by growers in highland or temperate zones. Unfortunately, the varieties available are seldom ideally suited to the local growing conditions. Hence, significant expansion of potato production in warm lowland areas usually requires a combination of improvements in seed technology, varieties, pest management, and storage.

Mediterranean and subtropical potato-growing systems are intermediate between those of temperate and tropical lowland zones in many respects. Farmers may plant potatoes in spring, summer, or fall. In North Africa and the Middle East, seed tubers for the spring and summer crops are imported from Europe; part of those tubers harvested in the spring crop are kept to use as seed in the fall crop. Some potatoes from the fall crop are exported to Europe in springtime, and potatoes are the principal vegetable crop export of the region.

Who eats potatoes and in what quantities?

Figures on potato consumption are usually derived from food balance sheets as the residual after net exports, nonfood uses (livestock feed, industry, seed), and waste have been subtracted from total production. All the components of these calculations are estimates, and the compounding of errors make consumption estimates highly suspect. Household surveys that have focused specifically on the issue of potato consumption indicate that, in general, the food balance sheets substantially underestimate potato consumption. In addition, as national averages, potato consumption estimates mask important differences in consumption among groups within a country.

Within rural potato-growing areas, producers eat more potatoes than nonproducers. Both of these groups eat more potatoes than rural people in other areas, because potatoes marketed outside of producing areas are usually consumed in large towns and cities. High transport costs and the low purchasing power of rural consumers discourage market agents from shipping potatoes from producing areas to rural deficit areas.

In towns and cities, the effect of household income on potato consumption depends on the relative price of potatoes. In the few developing areas where potatoes are relatively cheap, such as in the Andes, potato consumption is highest among poor households. In the more common situation where potatoes are relatively expensive, as in most of Africa and Asia, potato consumption is highest among the wealthy. The implication is that as income levels increase in the future, so will potato consumption in most places.

What nutrients do potatoes provide in the human diet?

The potato contributes not only energy but also substantial amounts of high-quality protein and essential vitamins, minerals, and trace elements to the diet. The energy and protein contents of fresh potatoes are much lower than those of cereals, but the differences are narrowed when these foods are cooked. The biological value of potato protein is better than that of most other vegetable sources and is comparable to that of cow's milk. Its high lysine content makes potato protein a valuable complement to cereal-based diets that are generally limiting in this amino acid. The potato is well balanced in the sense that the protein-to-calorie ratio is higher than that of other root crops, most cereals, and plantains. However, if one satisfies one's entire protein requirement with potatoes, one could still be short of calories. This illustrates the error of considering the potato primarily as a source of starch or calories.

The potato is comparable to other common vegetables in vitamin content and is especially rich in vitamin C. As little as 200 grams of boiled potato provides an adult's recommended daily allowance of vitamin C. A potato's mineral content is strongly influenced by the soil in which it is grown. Normally, the potato is a moderately good source of iron, a good source of phosphorus and magnesium, and an excellent source of potassium.

These nutritional facts indicate that the potato's contribution to the diet is not principally energy but rather protein, vitamins, and minerals.

How do price and income changes affect the demand for potatoes?

The demand for potatoes depends on population size, income levels, prices, and food habits. Conventional projections of demand assume that changes in income have only a small effect on demand. In other words, it is widely assumed that the "income elasticity of demand" for potatoes is low. It is also assumed that changing prices and food habits have little impact on the demand for potatoes. But recent studies show that in most developing countries, potatoes are a luxury, not a staple food, and, hence, the income elasticity of demand is high. In many places, potato prices are falling and food habits are changing to include more potatoes in the diet. As a result, the published demand projections for potatoes have fallen short of actual increases in demand. As noted earlier, potato production has increased faster than production of most other crops. Most of the additional potatoes were for human consumption and not for export or for livestock feed. In other words, the demand for potatoes has grown more rapidly than the demand for most foods in developing countries. As average potato consumption is still less than a quarter of that in Western Europe, it is likely that the demand for potatoes will continue to grow rapidly in most developing areas.

How can price controls, storage, processing, and foreign trade stabilize prices or expand the market for potatoes?

A few governments have attempted to control potato prices or dampen their fluctuations by buying and storing potatoes at harvest time when they are cheap and selling them later when they are scarce. The idea makes sense, but, in practice, these programs have been short-lived because the price movements were difficult to predict and storage costs and losses were higher than expected.

Although most large-scale government-operated storage schemes have failed, potato farmers practically everywhere store a part of their harvest. Farm storage helps to stabilize market supplies and prices throughout the year. Farm storage of seed tubers can also help smooth the flow of potatoes to consumer markets by allowing farmers to grow potatoes at different times of the year.

Manufacturing potato starch and alcohol is seldom practical in developing countries because potatoes are too expensive relative to other raw materials. There are a few exceptions, however, such as in northern China and southern Chile, where potatoes are cheap at harvest time and the costs of transporting them to urban markets are high. Simple potato processing for human consumption is generally more promising than starch or alcohol production.

Over 98 percent of the potatoes grown in developing countries are consumed domestically rather than exported. In North Africa and the Middle East, a small volume of potato trade generates significant returns in foreign exchange. But for most nations, trade in potatoes does not have great promise.

Why should scarce public resources be allocated to potato research and extension?

Public funding is needed because private firms have little reason to conduct potato research and extension. Few of the benefits of agricultural research and extension can be recovered by private enterprises because they accrue to all members of society through lower production costs and food prices. Most improvements in potato production cannot be patented and marketed at a profit, so there is little incentive for private firms to develop them. Because most potatoes are grown by small farmers and consumed domestically in fresh form, neither the farmers, processors, nor international trading interests can be expected to finance potato research and extension, as they do for export crops like sugar, tea, and jute.

What are the essential ingredients of a successful potato program?

The backlog of scientific information on potatoes in developed countries together with practically-oriented research promoted by the International Potato Center (CIP) and regional networks has created many opportunities for high returns to national potato programs in developing countries. But not all programs are successful.

The optimal size and structure of a potato program depend on the country's level of development, the institutional framework, and the personnel and physical and financial resources available for potato improvement. Programs with foreign funding that attempt to minimize bureaucratic problems by working outside of normal administrative channels generally collapse when the external support is withdrawn. Therefore, strong local commitment and integration into the existing agricultural research and development system are essential.

A national program needs a solid research base to be able to utilize and adapt the results of research conducted at the international agricultural research centers and other institutions around the world. Most national programs should have some expertise in four key areas: seed systems, varieties, pest management, and storage. The specific priorities within each of these areas should result from a careful examination of local production problems and the program's resources.

Detailed information on the main constraints to potato production and use is needed in order to set appropriate goals for technological research and extension. A number of programs that have generated an impressive amount of new information and technology have had negligible impact on production or social welfare because they worked on topics that had little practical importance or produced technologies that were impractical for the typical farmer.

Interaction among researchers, extension agents, and their clients (farmers, market agents, consumers) is essential for identifying and solving problems. Communication can be facilitated by interdisciplinary team research that involves both technologists and social scientists. Farmers are usually eager to participate in research on their farms, as long as it has a practical short-term payoff.

Once appropriate priorities have been set, the effectiveness of research depends on the analytical capabilities of researchers and the information base from which they draw. Both can be enhanced through contact with other scientists and access to scientific literature generated by other institutions. CIP can help national programs tap various information, training, and technical assistance sources.

National programs that need financial resources and resident foreign personnel can benefit from direct associations with bilateral agencies. Increasingly, these agencies and CIP are working together to eliminate duplication, improve coordination, and provide greater continuity to potato improvement programs.

What impact have successful potato programs had?

A review of successful potato programs in India, Rwanda, Tunisia, Vietnam, Sri Lanka, and Central America reveals one factor in common: the high priority given to seed production and distribution. Because of differences among the countries, and among regions within some countries, each program had to develop its own solutions to the seed problem.

In the two studies where economic costs and benefits were estimated (Tunisia and Rwanda), the potato programs' rates of return were substantially higher than those of most other development projects.

Aside from yield increase, which is a central measure in most program evaluations, successful potato programs have had several other types of impact. For example, the Rwandese seed program facilitated the rapid introduction of new, blight-resistant varieties that can be grown in the rainy season. The Indian program contributed to a massive expansion of the area planted with potatoes on the plains in the winter season. The increase in planting area in turn led to intensification of the cropping system, increased employment and rural incomes, substantial declines in potato prices, and increased consumption. Improved seed storage in Sri Lanka helped to cut post-harvest losses, reduce costs, extend the potato-growing season, and reduce dependence on imported seed.

What are the prospects for potatoes in developing countries?

In most developing countries, potato production is expanding rapidly. In many places where potatoes are still little known or eaten only occasionally as a vegetable, they are likely to become an important vegetable in the diet. In some areas where potatoes are now consumed as a major vegetable, they may become a staple food. In those areas where potatoes are already a staple food (for example the Andes and temperate South America), the rate of growth of production will probably be slower than in other areas. Hence, per capita potato consumption can be expected to rise in most, but not all, regions. Policymakers and researchers can best tap the unexploited potential of the potato as a food crop by implementing sound research and extension programs.

A sharp increase in supply, coupled with storage and transportation problems, may lead to temporary market surpluses and low prices. But if the enlarged supply is sustained over time by new cost-reducing technology, this will encourage consumers to eat more potatoes and make potato production profitable even at lower prices. Home economists, school teachers, and rural development workers can help stimulate higher levels of potato consumption by spreading information on the nutritional value of potatoes as well as different ways to prepare them. Nonetheless, the key to expanding production and consumption is not promotional campaigns but effective productivity-increasing programs.

The central goal of a potato program should be to identify and solve production and marketing problems that make potatoes scarce and expensive. The specific problems that merit priority attention vary from place to place. This highlights the need for strong client-oriented national research programs that have both technical and socioeconomic expertise.

References

This bibliography lists some of the most important references on potatoes in developing countries as well as more general studies on agricultural production, marketing, and programs. Detailed technical publications on potatoes are available from the International Potato Center, Aptdo 5969, Lima, Peru.

Ahmad, K. U. 1977. *Potatoes for the tropics.* Dacca: Mumtaj Kamal.

Arndt, T. M.; Dalrymple, D. G.; and Ruttan, V. W., eds., 1977. *Resource allocation and productivity in national and international agricultural research.* Minneapolis: University of Minnesota Press.

Arnon, I. 1975. *The planning and programming of agricultural research.* Rome: FAO.

Beukema, H. P., and van der Zaag, D. E. 1979. *Potato improvement.* Wageningen: International Agricultural Centre.

Booth, R. H., and Shaw, R. L. 1981. *Principles of potato storage.* Lima: International Potato Center.

Boyce, J. K., and Evenson, R. E. 1975. *Agricultural research and extension programs.* New York: Agricultural Development Council.

Brush, S.; Carney, H. J.; and Huaman, Z. 1981. Dynamics of Andean potato agriculture. *Economic Botany* 35:70–88.

Burton, W. G. 1966. *The potato: A survey of its history and of factors influencing its yield, nutritive value, quality and storage.* Wageningen: H. Veenman and Zonen.

Calkins, P. H., and Su-hua Tu. 1978. White potato production in Taiwan. *AVRDC (Asian Vegetable Research and Development Center) Technical Bulletin* 10:78–87.

CGIAR (Consultative Group on International Agricultural Research). 1980. *Consultative Group on International Agricultural Research.* Washington, D.C.

______. 1984. *Report of the second quinquennial review of the Centro Internacional de la Papa.* Rome: FAO.

______. 1985. *Summary of international agricultural research centers: A study of achievements and potential.* Washington, D.C.

Chatha, I. S., and Sidhu, D. S. 1980. *Production and marketing of potato in the Punjab state.* Ludhiana: Punjab Agricultural University, Department of Economics and Sociology.

Dalrymple, D., and Akeley, R. V. 1968. *The potato industry in East Pakistan.* Washington, D.C.: U.S. Department of Agriculture, International Agricultural Development Service.

Daniels, D., and Nestel, B., eds. 1981. *Resource allocation to agricultural research.* Ottawa: International Development Research Centre.

Dürr, G. 1980. Potato production and utilization in Kenya. Ph.D. diss., Technical University Berlin.

______. 1983. *Potato production and utilization in Rwanda.* Social Science Department Working Paper 1983-1. Lima: International Potato Center.

Dürr, G., and Lorenzl, G. 1980. *Potato production and utilization in Kenya.* Lima: International Potato Center.

Eicher, Carl K., and Staatz J. M. 1984. *Agricultural development in the Third World.* Baltimore: Johns Hopkins University Press.

Elias, S. M., and Islam, N. M. 1982. *Socio-economic assessment of improved technology of potato and identification of constraints to its higher production.* BARI Agricultural Economics Research Report No. 82-4. Joydebpur: Bangladesh Agricultural Research Institute.

Espinoza, N. O.; Estrada, R.; Silva-Rodriguez, D.; Tovar, P.; Lizarraga R.; and Dodds, J. H. 1986. The potato: A model crop plant for tissue culture. *Outlook on Agriculture* 15:21–26.

Fano, H. 1983. Cambio tecnológico y tendencias de la producción de papa en la región central del Perú 1948–1979. Thesis, Universidad Nacional Agraria, La Molina, Peru.

Ferroni, M. 1985. *La economía de la papa en México.* Mexico City: Instituto Nacional de Investigaciones Agrícolas, Programa Regional Cooperativo de Papa.

Fishel, W. L., ed. 1971. *Resource allocation in agricultural research.* Minneapolis: University of Minnesota Press.

Food and Agriculture Organization (FAO). 1971. *Agricultural commodity projections,* 1970–80. Rome: FAO.

Franco, E.; Horton, D.; Cortbaoui, R.; Tardieu, F.; and Tomassini, L. 1980. *Evaluación agroeconómica de ensayos conducidos en campos de agricultores en el Valle del Mantaro (Perú) Campaña 1978/79.* Social Science Department Working Paper 1980–4. Lima: International Potato Center.

Franco, E.; Horton, D.; Cortbaoui, R.; Tomassini, L.; and Tardieu, F. 1981. *Evaluación agroeconómica de ensayos conducidos en campos de agricultores en el Valle del Mantaro (Perú). Campaña 1979/80.* Social Science Department Working Paper 1981-1. Lima: International Potato Center.

Franco, E.; Horton, D.; and Tardieu, F. 1979. *Producción y utilización de la papa en el Valle del Mantaro, Perú.* Social Science Department Working Paper 1979-1. Lima: International Potato Center.

Franco, E.; Moreno, C.; Alarcón, J. 1983. *Producción y utilización de la papa en la región del Cuzco. Resultados de una encuesta de visita única.* Social Science Department Working Paper 1983-2. Lima: International Potato Center.

Franco, E., and Schmidt, E. 1984. *Adopción y difusión de variedades de papa en el departamento de Cajamarca.* Social Science Department Working Paper 1985-1. Lima: International Potato Center.

Fu, G. 1979. *Producción y utilización de la papa en Chile.* Lima: International Potato Center.

Gray, R. W.; Sorenson, V. L.; and Cochrane, W. W. 1954. *An economic analysis of the impact of government programs on the potato industry of the United States.* North Central Regional Publication No. 42. Minneapolis: University of Minnesota Agricultural Experiment Station.

Harris, P. M., ed. 1978. *The potato crop.* London: Chapman and Hall.

Hawkes, J. G. 1978a. Biosystematics of the potato. In *The potato crop,* ed. P. M. Harris, 15–69. London: Chapman and Hall.

______. 1978b. History of the potato. In *The potato crop,* ed. P. M. Harris, 1–14. London: Chapman and Hall.

Hayami, Y., and Ruttan, V. W. 1985. *Agricultural development.* Baltimore: Johns Hopkins University Press.

Hooker, W. J., ed. 1983. *Research for the potato in the year 2000.* Lima: International Potato Center.

Horton, D. 1981. A plea for the potato. *Ceres* 14(1):28–32.

______. 1983. Potato farming in the Andes. *Agricultural Systems* 12:171–184.

______. 1984. *Social scientists in agricultural research.* Ottawa: International Development Research Centre.

______. 1986. Assessing the impact of international agricultural research and development programs. *World Development* 14:453–468.

Horton, D.; Cortbaoui, R.; Hatteb, H.; and Monares, A. In press. Assessing the impact of an agricultural development project: Seed potato multiplication in Tunisia. *Quarterly Journal of International Agriculture.*

Horton, D., and Fano, H. 1985. *Potato atlas.* Lima: International Potato Center.

Horton, D.; Lynam, J.; and Knipscheer, H. 1984. Root crops in developing countries: An economic appraisal. In *Proceedings of the sixth symposium of the International Society for Tropical Root Crops.* Lima: International Potato Center.

Horton, D.; Tardieu, F.; Benavides, M.; Tomassini, L.; and Accatino, P. 1980. *Tecnología de la producción de papa en el Valle del Mantaro, Perú.* Social Science Department Working Paper 1980-1. Lima: International Potato Center.

Howard, H. W. 1978. The production of new varieties. In *The potato crop,* ed. P. M. Harris, 607–646. London: Chapman and Hall.

Huamán, Z. 1986. *Systematic botany and morphology of the potato.* Technical Information Bulletin 6. Lima: International Potato Center.

Huamán, Z., and Ross, R. W. 1985. Updated list of potato species names, abbreviations, and taxonomic status. *American Potato Journal* 62:629–641.

International Potato Center. 1984. *Potatoes for the developing world.* Lima.

Jaynes, J. M.; Espinoza, N.; and Dodds, J. H. In press. Genetic engineering of the potato. *Scientific American.*

Johnston, B. F., and Kilby, P. 1975. *Agriculture and structural transformation.* New York: Oxford University Press.

Jones, W. O. 1972. *Marketing staple food crops in tropical Africa.* Ithaca: Cornell University Press.

______. 1984. Economic tasks for food marketing boards in tropical Africa. *Food Research Institute Studies* 19:113–138.

Laufer, B. 1938. *The American plant migration. Part I: The potato.* Anthropological Series. Chicago: Field Museum of Natural History.

Li, P. H., ed. 1985. *Potato physiology.* Orlando: Academic Press.

Mante, W., and Blodig, W. 1979. *Bibliography on socio-economic aspects of potato production and utilization.* Berlin: Technical University.

Mayer, E. 1979. *Land use in the Andes.* Lima: International Potato Center.

Mellor, J. W. 1966. *The economics of agricultural development.* Ithaca: Cornell University Press.

Mendoza, H. A., and Sawyer, R. L. 1985. The breeding program at the International Potato Center. In *Progress in plant breeding,* ed. G. E. Russell, 1:117–137. London: Butterworths.

Miah, M. 1984. Potato cultivation in Bangladesh. University of Dhaka. Mimeo.

Mokyr, J. 1983. *Why Ireland starved.* London: Allen & Unwin.

Monares, A. 1981. The potato seed system in the Andean region. Ph.D. diss., Cornell University, Ithaca, N.Y.

______. 1984. *Building an effective potato country program: The case of Rwanda.* Social Science Department Working Paper 1984-3. Lima: International Potato Center.

Montaldo, A. 1964. *Bibliografía Latinoamericana sobre papa.* Maracay: Universidad Central de Venezuela, Facultad de Agronomía.

Nagaich, B. B.; Shekhawat, G. S.; Guar, P. C.; and Verma, S. C., eds. 1982. *Potato in developing countries.* Delhi: Indian Potato Association.

Niñez, V. K. 1984. *Household gardens.* Potatoes in Food Systems Research Series Report 1. Lima: International Potato Center.

Pinstrup-Andersen, P. 1982. *Agricultural research and technology in economic development.* New York: Longman.

Poats, S. 1981. La pomme de terre au Rwanda: Résultats préliminaires d'une enquête de consommation. *Bulletin Agricole du Rwanda* Mai:82–91.

______. 1983. Beyond the farmer: Potato consumption in the tropics. In *Research for the potato in the year 2000,* ed. W. J. Hooker, 10–17. Lima: International Potato Center.

Potts, M., ed. 1983. *On-farm potato research in the Philippines.* Lima: International Potato Center.

Pushkarnath. 1976. *Potato in subtropics.* Delhi: Orient Longman.

Rhoades, R. 1982a. Toward an understanding of hot, humid tropical farming systems with emphasis on the potato. In *Potato production in the humid tropics,* eds. L. J. Harmsworth, J.A.T. Woodford, and M. E. Marvel, 444–455. Los Baños (Philippines): International Potato Center Region VII.

______. 1982b. The incredible potato. *National Geographic* 161:668–694.

______. 1984. *Changing a post-harvest system: Impact of diffused light potato stores in Sri Lanka.* Social Science Department Working Paper 1984-1. Lima: International Potato Center.

———. 1985. *Breaking new ground: Agricultural anthropology.* Lima: International Potato Center.

———. 1986. *Potato production zones and systems of developing countries.* Lima: International Potato Center.

Rhoades, R., and Booth, R. 1982. Farmer-back-to-farmer: A model for generating acceptable agricultural technology. *Agricultural Administration* 11:127–137.

Rhoades, R.; Booth, R.; and Potts, M. 1983. Farmer acceptance of improved potato storage practices in developing countries. *Outlook on Agriculture* 12(1):12–16.

Ruthenberg, H. 1980. *Farming systems in the tropics.* 3rd ed. Oxford: Oxford University Press.

Ruttan, V. W. 1982. *Agricultural research policy.* Minneapolis: University of Minnesota Press.

Salaman, R. N. 1986. *The history and social influence of the potato.* rev. ed. Cambridge: Cambridge University Press.

Sawyer, R. 1982. *Profile: the International Potato Center 1970–2000.* Lima: International Potato Center.

Scobie, G. M. 1979. *Investment in international agricultural research.* Staff Working Paper No. 361. World Bank: Washington, D.C.

———. 1984. *Investment in agricultural research: Some economic principles.* Working paper. Mexico City: International Maize and Wheat Improvement Center.

Scott, G. 1981. Potato production and marketing in Central Peru. Ph.D. diss., University of Wisconsin, Madison.

———. 1983. *Marketing Bhutan's potatoes.* Lima: International Potato Center.

———. 1985. *Markets, myths and middlemen: A study of potato marketing in Central Peru.* Lima: International Potato Center.

———. 1986a. *Marketing Thailand's potatoes.* Lima: International Potato Center.

———. 1986b. *La pomme de terre en Afrique Centrale.* Lima: International Potato Center.

Scott, G., and Costello, G., eds. 1985. *Comercialización interna de alimentos nacionales en América Latina.* Ottawa: International Development Research Centre.

Shaw, R., and Booth, R. 1982. *Simple processing of dehydrated potatoes and potato starch.* Lima: International Potato Center.

Shumway, C. R. 1983. Ex ante research evolution: Can it be improved? *Agricultural Administration* 12:91–102.

Simmonds, N. W. 1979. *Principles of crop improvement.* New York: Longman.

Smith, O. 1977. *Potatoes: Production, storing, processing.* 2nd ed. Westport: AVI Publishing.

Somaratne, S. M. 1985. Generation, adoption and impact of diffused light potato storage technology in Badulla District, Sri Lanka. Master's thesis, University of the Philippines, Los Baños.

Srivastava, B. N. 1980. *Potato in the Indian economy.* Social Science Department Working Paper 1980-2. Lima: International Potato Center.

Stone, B. 1984. An analysis of Chinese data on root and tuber crop production. *The China Quarterly* 99:594–630.

Thornton, R. E.; and Sieczka, J. B., eds. 1980. *Commercial potato production in North America.* American Potato Journal Supplement, vol. 57.

Timmer, C. P.; Falcon, W. P.; and Pearson, S. R. 1983. *Food policy analysis.* Baltimore: Johns Hopkins University Press.

Ugent, D. 1970. The potato: What is the botanical origin of this important crop plant and how did it first become domesticated? *Science* 170:1161–1166.

Uyen, N. V. and van der Zaag, P. 1983. Vietnamese farmers use tissue culture for commercial potato production. *American Potato Journal* 60:873–879.

______. 1985. Potato production using tissue culture in Vietnam: The status after four years. *American Potato Journal* 62:237–241.

Valderrama, M., and Luzuriaga, H. 1980. *Producción y utilización de la papa en el Ecuador.* Lima: International Potato Center.

van der Zaag, D. E. 1983. The world potato crop: The present position and probable future developments. *Outlook on Agriculture* 12(2):63–72.

van der Zaag, D. E., and Horton, D. 1983. Potato production and utilization in world perspective with special reference to the tropics and sub-tropics. In *Research for the Potato in the Year 2000,* ed. W. J. Hooker, 44–58. Lima: International Potato Center.

Vargas, D. 1983. Análisis económico de algunos factores en la producción de papa: Caso del valle de Cañete. Master's thesis, Universidad Nacional Agraria, La Molina, Peru.

Werge, R. 1977a. *Potato storage systems in the Mantaro Valley region of Peru.* Lima: International Potato Center.

______. 1977b. *Socioeconomic aspects of the production and utilization of potatoes in Peru: A bibliography.* Lima: International Potato Center.

———. 1979a. *The agricultural strategy of rural households in three ecological zones of the central Andes.* Social Science Department Working Paper 1979-4. Lima: International Potato Center.

———. 1979b. Potato processing in central highlands of Peru. *Ecology of Food and Nutrition* 7:229–234.

Woolfe, J. A. 1987. *The potato in the human diet.* Cambridge: Cambridge University Press.

Young, N. A. 1981. *The European potato industry.* Ashford (England): Centre for European Agricultural Studies.

Index

About the Book and Author

Although the potato is usually thought of as a temperate-zone crop, potato growing in the tropics and subtropics is spreading rapidly. In terms of the dollar value of the crop, this edible root now ranks fourth in the developing world after rice, wheat, and maize. Nevertheless, policymakers often underrate the importance of the potato as a source of employment, income, and food or they underestimate the potential benefits from expanding potato production and use. The payoff from applied research on potato cultivation in the tropics and subtropics is high due to the large body of scientific information from developed countries.

This book summarizes the principles of potato production, distribution, and use. The essential facts about the potato as a crop, a commodity, and a food are discussed as well as the issues that scientists and policymakers should consider in setting priorities for implementing and assessing the impact of potato research and extension programs. A major premise of the book is that programs aiming to increase food supplies and reduce poverty through crop improvement need to consider not only production technology but also marketing strategies and consumption patterns. Adequate planning for agricultural research and development requires an understanding of how crops are grown, marketed, and used and of what potential benefits the new technologies can yield. Hence, effective crop improvement programs need both technical and socioeconomic expertise. The administrators and others responsible for implementing these programs must concern themselves with the policies that impinge on the adoption and consequences of new production methods so that their countries may reap the full benefits of an increased and stable food supply.

Douglas Horton is the head of the Social Science Department, International Potato Center, Lima, Peru.